An Introduction to Forensic Biology

WHEN THE BODY SPEAKS

FIRST EDITION

Gilbert Ellis
Barry University

cognella® | ACADEMIC PUBLISHING

Bassim Hamadeh, CEO and Publisher
Kassie Graves, Director of Acquisitions
Jamie Giganti, Senior Managing Editor
Miguel Macias, Graphic Designer
Jennifer McCarthy, Acquisitions Editor
Sean Adams, Project Editor
Luiz Ferreira, Senior Licensing Specialist
Kat Ragudos, Interior Designer
Abbey Hastings, Associate Production Editor

Cover image copyright © Depositphotos/clearviewstock.
          copyright © Depositphotos/paulrommer.
          copyright © Depositphotos/jochenschneider.
          copyright © Depositphotos/paulfleet.
          copyright © Depositphotos/eddiephotography.

Printed in the United States of America

ISBN: 978-1-5165-1244-7 (pbk) / 978-1-5165-1245-4 (br)

cognella® | ACADEMIC PUBLISHING

CONTENTS

Dedication   vi
Acknowledgments   vii

Chapter 1: Organ Systems, Pathology, and Autopsy   1

Learning Objectives   1
For Further Reading   9
Review questions for Chapter 1: Organ Systems, Pathology, and Autopsy   11

Chapter 2: Body Fluids: Serology   13

Learning Objectives   13
For Further Reading   23
Review question for Chapter 2: Body Fluids: Serology   25

Chapter 3: Blood Spatter Pattern Analysis   27

Learning Objectives   27
Review questions for Chapter 3: Blood Spatter Pattern Analysis   39

Chapter 4: Forensic Anthropology   41

Learning Objectives   41
For Further Reading   56
Review questions for Chapter 4: Forensic Anthropology   57

Chapter 5: Forensic Odontology   59

Learning Objectives   59
For Further Reading   65
Review questions for Chapter 5: Forensic Odontology   67

Chapter 6: Dermatoglyphics: Skin, Hair, Prints, Fibers, and Burns   69

Learning Objectives   69

For Further Reading   80

Review questions for Chapter 6: Dermatoglyphics: Skin, Hair, Prints,
Fibers, and Burns   81

Chapter 7: Forensic DNA Analysis   83

Learning Objectives   83

For Further Reading   90

Review questions for Chapter 7: Forensic DNA Analysis   91

Chapter 8: Forensic Toxicology, Alcohol, and Drugs   93

Learning Objectives   93

For Further Reading   106

Review questions for Chapter 8: Forensic Toxicology, Alcohol, and Drugs   107

Chapter 9: Forensic Entomology   109

Learning Objectives   109

For Further Reading   119

Review questions for Chapter 9: Forensic Entomology   121

Chapter 10: Forensic Botany   123

Learning Objectives   123

For Further Reading   130

Review questions for Chapter 10: Forensic Botany   133

References   135

Glossary   139

Appendix A   145

Appendix B   153

Appendix C   155

Credits   161

Dedication

The greatest gift that I have had during my long tenure in academics is the strong support of my family. To that end I dedicate this writing to my wife, Patricia, my daughter, Cassandra, and my son, Christopher, as well as their families. Their continued support exceeds all of my expectations. With love, Dad.

Acknowledgments

During the period I have been involved with writing this book, I have had the support of many people within my family as well as my colleagues at work. I especially want to thank Ms. Yaniet Hinson for her dedication to capturing and recording the many images within the text. Her support and effort has been extraordinary. I thank Ms. Lisa Dougherty for her help in assembling the glossary and working on gathering certain images. Additionally, Luis Jarvis and my colleagues Dr. Alan Sanborn and Dr. Gerhild Packert provided professional photographs used as images in the book. I thank Barry University for allowing me a sabbatical leave during the fall of 2013 to complete the writing. I also thank the staff at Cognella Academic Publishing for supporting me in the publication and marketing of the text.

ORGAN SYSTEMS, PATHOLOGY, AND AUTOPSY

LEARNING OBJECTIVES

After completion of this chapter, students will be able to:

- » Identify the major organ systems in the human body
- » Explain how each organ system relates to the forensic autopsy
- » Interpret the cause, manner, and mode of death in a forensic case
- » Distinguish the major types of pathological wounds in homicide cases
- » Compare the postmortem changes such as rigor mortis, algor mortis, and livor mortis
- » Design a protocol for a forensic autopsy

The forensic pathologist is a highly skilled medical doctor who has been trained in clinical and anatomical pathology. The word pathology refers to the study and understanding of the diseased (abnormal) states of the various organ systems in the human body. A basic understanding of the organization of the human body is essential in order to study the pathological condition of each organ. The following information provides a brief review of the organization of the human body and the morphology and physiology of each of its organ systems. There are eleven organ systems which include the integumentary, skeletal, muscular, nervous, endocrine, cardiovascular, lymphatic, respiratory, digestive, urinary and reproductive. The hierarchy upon which these organ systems develop follows the basic building block concept of all living organisms. These levels of organization include: chemical consisting of small and large molecules, cells, tissues, organs, organ systems and finally the organism. The eleven organ systems which are found in man (as well as many other vertebrate organisms) are briefly described in the inserts in figure 1.1. Within each of these organ systems the pathologist needs to be familiar with all of the details of the morphology (structure) and physiology (function). The anatomical pathologist is an expert at determining the pathological condition of every organ in the human body. Much of the work that is performed by the anatomical pathologist is done by making microscopic observations of the various tissues from the organs within the human body. Often the determination of whether or not the organ is diseased may be documented by observing the cellular structure of a particular organ. The forensic pathologist works closely with the cytotechnology laboratory whose responsibility is to prepare microscopic sections of the various organs so that they may

Human Body Systems

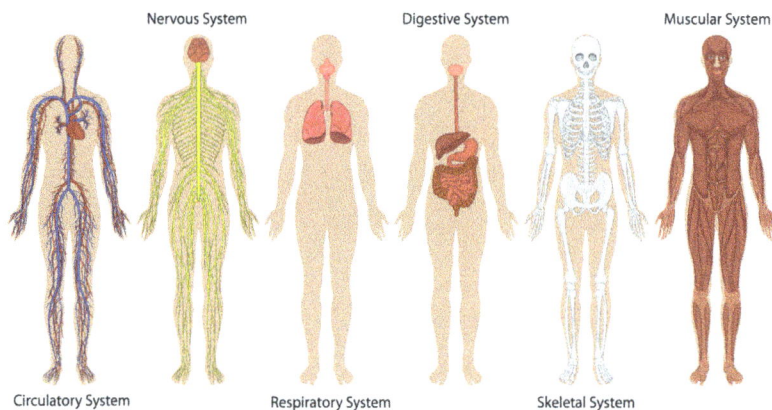

FIGURE 1.1
Organ Systems.

be observed and analyzed under the microscope. From the forensic pathologist's point of view it is important to determine whether or not the tissues in the body have changed due to a natural diseased state or whether the tissue has developed abnormal morphology due to an unnatural event.

Forensic pathology involves postmortem investigation of sudden or unexpected death. In most states in the United States the law specifies that any unnatural death requires a forensic pathologist to conduct an autopsy. The purpose of the forensic autopsy falls under three broad determining factors related to establishing the cause, manner and mode of death (Table 1.1).

There are a series of natural postmortem changes which occur after death and they include: rigor mortis, algor mortis, livor mortis, desiccation, putrefaction and autolysis. As a result of these and other changes (gastric emptying, changes in chemical compositions of body fluids and tissues, and insect activity) that occur post mortem the pathologist is able to make a determination as to the time of death which is of critical importance to reconstruct the events of the particular case.

Table 1.1 Determining factors of forensic autopsy.

1. Establish the cause of death
a. Asphyxiation
b. A blow to the head
c. An infection
2. Determine the manner of Death
a. Instrument causing the death
b. Illness causing the death
3. Determine the mode of death
a. Natural
b. Suicide
c. Accident
d. Homicide

Rigor Mortis

Skeletal muscles continue producing ATP by glycolysis for a short time after death. The ATP is hydrolyzed to ADP and lactic acid accumulates in the absence of oxygen. As a result there is a buildup of ADP and an absence of ATP which in an environment of increased calcium ions and a decreased pH the actin-myosin crossbridge formation is unable to be released which causes the muscles to go into a somewhat permanent state of contraction (rigidity). The onset of rigor begins between ½ to 1 hour after death and proceeds to a complete tetany between 12 to 24 hours and then proceeds to lyse between 24 – 36 hours so that rigor is not present after 36 hours in a normal environment. Table 1.2 explain conditions which will either accelerate or decelerate the time period for the cycle of rigor.

Table 1. 2 **Conditions which accelerate/decelerate rigor.**

Accelerate	Decelerate
Prior exercise	Cold environmental temperature
Acidosis	Hypothermia
Seizures	Decreased muscle mass
Electrocution	Drugs & poisons (CO)
Hyperpyrexia	
Hot environmental temperatures	
Drugs (i.e. strychnine)	
Cadaveric spasm	

Algor Mortis

Algor mortis or post mortem cooling has been extensively investigated. There is a wide margin of error even in controlled conditions. There are a large number of factors that can produce the high degree of variability in temperature change after death. The first of these is that one cannot assume that the temperature at the time of death was normal due to a multitude of reasons such as individuals having increased body temperatures due to infection or having lowered body temperatures due to some possible metabolic disorder (hypothyroid). In a stable environment body temperature declines until it reaches ambient temperature but it may not decline at a constant rate. A rule of thumb calculation would be to 98.4 degrees F subtract the rectal temperature and divide by 1.5 which will equal the hours since death. For example a rectal temperature recorded at 86.4 degrees F would estimate the time since death to have been approximately 8 hours. The mechanisms by which heat is lost from a corpse involves conduction which is transfer of heat by direct contact with another surface, radiation which is transfer of heat to surrounding air by infrared waves and convection which involves transfer of heat in association of moving air currents adjacent to the body. There are also many factors which influence the rate of heat loss such as the temperature gradient, moving air (wind speed) clothing, body fat, surface area and the nature of the environment at the time of measurement.

Livor Mortis

This condition is the result of blood being a liquid tissue and the effects that gravity has on a liquid within a container. After death blood pools (sets) in dependent tissues throughout the body as a result of gravity acting on the blood. An individual who was in the supine position (lying on his back) at the time of death will demonstrate complete livor mortis within 8-12 hours after death which results in the skin turning a dark purple color in relation to the pooling of blood in the dorsal tissues. This occurs in most all body tissues except for pressure points (elbows, buttocks, back of the head, heels, knees) those locations which make close and constant contact with a particular surface. Tardieu's spots and petichiae are specific examples of the accumulation of small pools of blood in the throat and eyes respectively. In some homicides where the victim is asphyxiated through strangulation small blood vessels in the sclera of the eye rupture and produce the traditional dark blue spots called petichiae.

Gastric Emptying Time

After a meal is consumed the bolus of food that is propelled to the stomach by peristaltic waves remains in the stomach between to 2-4 hours. During this time the particulate matter in the bolus is mixed with gastric juices from the stomach and undergoes reduction to a material called chyme (fluid consistency of a milk shake). Gastric emptying begins approximately 4 hours after ingestion and it is during this time period that through careful observation of the stomach contents on post mortem examination that the forensic pathologist may estimate the time since the last meal. This observation may corroborate information as to when the individual was seen eating his last meal and whether this is consistent with the time the individual was found to be dead. Gastric contents also provide important information regarding where the person ate, possible drug or poison ingestion after laboratory tests are performed on the stomach contents.

Chemical Changes in Tissues and Body Fluids

Various body fluids like blood, spinal fluid, aqueous humor and vitreous humor of the eye demonstrate chemical changes immediately or shortly after death. These changes are measureable and proceed in a somewhat orderly manner until the body goes through late stages of decomposition. Each change has its own unique time factor or rate. Determination of these chemical changes could help the forensic pathologist to determine time since death more precisely. The calculations of these changes that may prove valuable include the potassium (K) content of the aqueous humor and lactic acid, ascorbic acid, non-protein nitrogen, sodium (Na), chloride (Cl), magnesium (Mg), and bicarbonate concentrations of the vitreous humor (Table 1.3).

Although no single measurement gives a completely reliable estimate of the postmortem interval, combinations of chemical assays can be useful supporting information in cases of un-witnessed death. Among these vitreous humor of the eye is relatively stable, less susceptible than other body fluids to rapid chemical changes and contamination, easily accessible and its composition is quite similar to that of aqueous fluid, cerebrospinal fluid and blood serum making it suitable for many analyses to estimate postmortem interval.

Table 1.3 **Mean potassium concentration in relation to time since death.**

TSD (hours)	Mean K⁺ (mEq/l) ± SD			
	Non-Burn Cases (176)		Burn Cases (24)	
	1st Eye	2nd Eye	1st Eye	2nd Eye
Within 12 hours	5.66 ± 1.76	6.78 ± 2.11	6.05 ± 1.59	7.20 ± 0.64
12.1 to 24 hours	8.59 ± 1.06	9.42 ± 1.14	9.84 ± 1.51	11.24 ± 1.34
Above 24 hours	12.88 ± 2.49	12.14 ± 1.98	N/D	15.00 ± 0.00

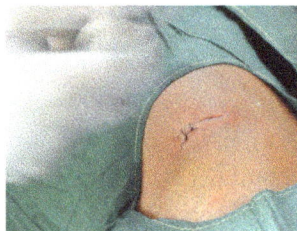
Incision

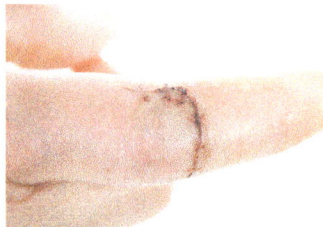
Laceration

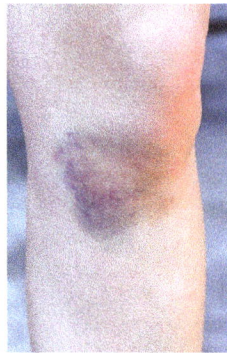
Contusion

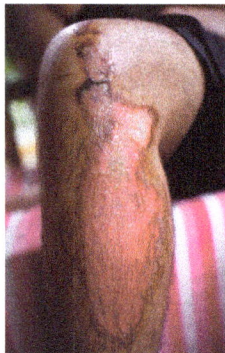
Abrasion

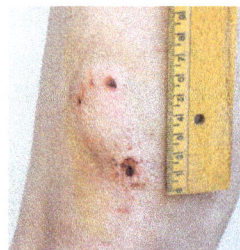
Puncture (Dog bite)

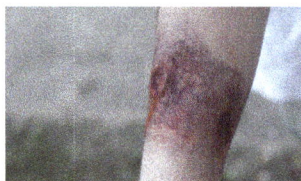
Entrance and Exit Wound (Gunshot)

FIGURE 1.2

a-f Wound types.

Pathology

The forensic pathologist is an expert in the identification and indication of any trauma that is present on the individual being examined as a result of an unnatural death. Determination of the type of wound that may be present is important in assessing means, manner and mode of death. The pathologist must first determine the type of wound present which could be a laceration, an incised wound, a puncture wound, an abrasion, a contusion or a gunshot wound (see figure 1.2 for examples).

Each of these types of trauma exhibit specific characteristics that must be described and documented. The dimensions of each wound, if present, must be measured to record length, width and depth. The position of the wound must be related to major anatomical landmarks on the body. A determination of the initial location must be identified if the wound involves cutting, slashing or laceration. Each wound must be reported in the autopsy summary with specific dimensions and located at the proper site on the body with the measurement of height above the heel. Wounds such as abrasions must be assessed to determine what type of surface caused the abrasion (rug, pavement, concrete, etc.). Contusions or bruises sometimes appear as a result of beating or striking an individual with a blunt object. Sub-surface color changes in contusions or bruises may be useful in determining time of injury. Table 1.4 outlines these color changes in relation to approximate time of injury.

Table 1.4 **Color changes in relation to time of injury.**

Dark blue/purple	1-18 hours*
Blue/brown	1-2 days*
Green	2-3 days*
Yellow	3-7 days*

*Assuming individual is in a good state of health

Gunshot Wounds – The proper identification of gunshot wounds allows the forensic pathologist to determine the type of firearm associated with the wound. In addition one of the most common determinations made by the forensic pathologist is the range of fire or the distance from the victim to the gun. Gunshot wounds are typically classified as; a.) contact, b.)intermediate range or c.) distant range. Contact wounds usually have soot on the outside of the skin and a muzzle imprint or laceration of the skin from the effects of gases released from the barrel of the weapon. Intermediate or close-range wounds may show a wide zone of powder stippling but lack a muzzle imprint and laceration. The area of powder stippling will depend upon the distance from the muzzle. Distant range wounds are lacking powder stippling and usually exhibit a hole roughly the caliber of the projectile fired. Entrance wounds into skull bone typically produce beveling or coning of the bone at the surface and are oriented away from the direction of the entrance wound. Exit wounds very often are irregular and may produce circumferential fractures that radiate outward from the skull plates at the site of exit. Exit wounds are generally larger than entrance wounds, due to the fact that the bullet has expanded or tumbled on its axis. Exit wounds do not exhibit gunshot residues or far less residues than those associated with entrance wound.

The Autopsy – An autopsy is an examination of a body after death. The forensic autopsy is used in an attempt to establish the cause of death, such as asphyxiation, a blow to the head, or an infection; the manner of death, which identifies the instrument or illness causing the death; and the mode of death, which is the circumstance of the death. There are four possible modes of death; natural death, accidental death, suicide, and homicide.

A forensic autopsy, or a medico-legal autopsy, is conducted by a forensic pathologist, chief medical examiner, or coroner, and is different from the more routine autopsy conducted in a hospital. The hospital autopsy is performed primarily to pinpoint the specific cause of someone's death when a natural cause is assumed. The forensic autopsy is performed to determine the cause of someone's death when the death was violent, sudden or suspicious. In addition to establishing the cause of death, the forensic autopsy often helps establish the time of death and identity of the deceased. A forensic autopsy also involves the collection and evaluation of evidence from the body and from the crime scene that can be used to implicate or exonerate a suspect and to support or refute an account of how the death actually occurred. A forensic autopsy requires not only the examination of a body in the facilities of the medical examiner but also the examination of evidence at the scene of the crime, including the circumstances leading up to and surrounding the death of an individual. An autopsy is always required in homicide cases and is common in cases in which there is suspected criminal violence, suicide, accident, sudden death, poisoning, when a body is to be cremated or dissected, and when death has occurred because of therapeutic procedures or from job-related activities. In this course we shall assume that any reference to an autopsy implies that the procedures being discussed follow those for a forensic autopsy.

The medico-legal officer, who conducts forensic autopsies, whether referred to as a medical examiner or as a coroner, must be not only a qualified doctor of medicine, but also a certified pathologist who is skilled in forensic pathology. Standards for his or her training, and for the procedures used in the conduction of autopsies, are established by the national Association of Medical Examiners.

In conducting a forensic autopsy, a specific procedure is followed. The forensic autopsy actually begins at the location where the body has been found. Before the body is moved, the scene where the body was found should be examined carefully for important evidence and people in the area should be interviewed. A substantial amount of information about the circumstances of death can be obtained at the scene of a homicide. This evidence can be destroyed or misleading evidence introduced by improper handling of the body or removal of the body from the scene. However, it is impossible to adequately examine a body at the location where it is found. Before it is removed from the scene to suitable facilities where a proper autopsy can be performed, the body should be moved as little as possible. Prior to transportation to the facilities of the medical examiner, the body should be wrapped in a clean, white sheet or placed in a clean body bag. Because of the risk of contamination, it would never be placed directly onto a cart, or in the back of an ambulance. At the morgue, the body should not be undressed until the clothing has been examined. Clothing, too, can yield evidence that can help to answer the questions addressed by the forensic autopsy. For the same reasons, the body should not be embalmed, nor should the fingerprints be lifted, before a complete examination of the clothing is completed.

The actual autopsy, the examination of the body, follows a prescribed procedure. The information discovered in each step of the procedure is documented and recorded so that all relevant information can be included in the written autopsy report, or protocol, prepared at the conclusion of the autopsy. The procedures for the autopsy on the fetal pig as well as a working and final autopsy report are included in the appendices of this book.

FOR FURTHER READING

Body of evidence: A radical new approach to forensic pathology. (2010, May 31). *Independent.* Retrieved from http://www.independent.co.uk/news/science/body-of-evidence-a-radical-new-approach-to-forensic-pathology-1987389.html

1. The organ system that is responsible for gas exchange (CO_2 and O_2) is the _____ system.
2. The condition that results in rigidity of the limbs after death is termed _____.
3. The purple discoloration of the skin due to pooling of the blood in dependent tissues is called _____.
4. The classic incision that is made for the forensic autopsy is called a(n) _____ incision.
5. The body plane that sections the body into equal right and left halves is the _____ plane.
6. The cause of death in a situation where an individual was suffocated with a pillow is that of _____.
7. Gastric emptying time is important because it gives the approximate time of the last meal. Typically, food is held in the stomach for _____ to _____ hours after death.
8. The organ system that is responsible for eliminating nitrogenous wastes from the body is the _____.
9. The four modes of death are _____, _____, _____, and _____.
10. A classic characteristic found in an individual who has been strangled is the presence of _____ in the sclera (white) of the eye.
11. The time for rigor mortis to set in is _____ if the person was electrocuted.
12. During an autopsy, the removal of the breast plate initially exposes the _____.
13. Bruises or _____ _____ are produced by rupturing the small blood vessels in the dermis of the skin.
14. Contact wounds produced from gunshots show evidence of _____ _____ around the edges of the wound as well as a muzzle print.
15. A wound produced by a sharp probe (such as an icepick) entering the body is classified as a _____ _____ wound.

BODY FLUIDS: SEROLOGY

LEARNING OBJECTIVES

After completion of this chapter, students will be able to:

- » Identify the major fluid compartments in the human body
- » Explain the major characteristics of the ABO blood grouping system
- » Apply the genetics of the ABO blood grouping system in a paternity case
- » Compare the major antigens in the ABO, Rh system
- » Choose the agglutination reaction that relates to the ABO, Rh blood groupings
- » Construct a diagram of the components of whole blood
- » Identify the major types of cells in human blood
- » Describe the significance of urine in the forensic investigation
- » Detect the presence of saliva and its significance to the forensic scientist
- » Interpret the findings of the PSA and acid phosphatase tests

Body fluids are very important indicators of the functioning of the organ systems. Abnormal metabolic states of the various organ systems in the human body are of great value to the forensic pathologist. Trying to determine cause and manner of death is very often dependent upon analyzing specific body fluids. The human body can be described in terms of fluid compartments as seen in the diagram below (figure 2.1).

The homeostatic regulation of fluid volumes within these compartments is of vital significance in maintaining normal metabolism. A loss or gain of fluid in any of the three major compartments will produce negative effects on the metabolic activity of the cells, tissues and organs related to the fluid values.

Based on a 70Kg. individual the largest of these fluid reservoirs is the intracellular fluid compartments (ICF) which consists of cytoplasm in the cells of the body and comprises roughly 30 liters in volume. The remaining fluid compartment is the extracellular fluid compartment (ECF) which consists of two separate reservoirs. One of these is the fluid surrounding the cells in the body, the interstitial fluid compartment (tissue fluid) which accounts for approximately 12 liters of fluid. The second and smallest of the fluid compartments is that of the blood plasma which accounts for approximately 3 liters of fluid contained within the vascular network. The chemistry of these fluid compartments can be studied by analyzing body fluids in the form of blood plasma, urine, saliva, cerebrospinal fluid, semen and possibly perspiration. Various metabolites found in these bodily fluids can be related to specific alterations of metabolism which might indicate poisoning or certain cellular characteristics of red and white cells from the blood and epithelial cells from the saliva and urine might be useful for identification of an individual through ABO blood groups or DNA analysis. We will examine each of these body fluids and discuss their nature and how they might be used in a forensic investigation.

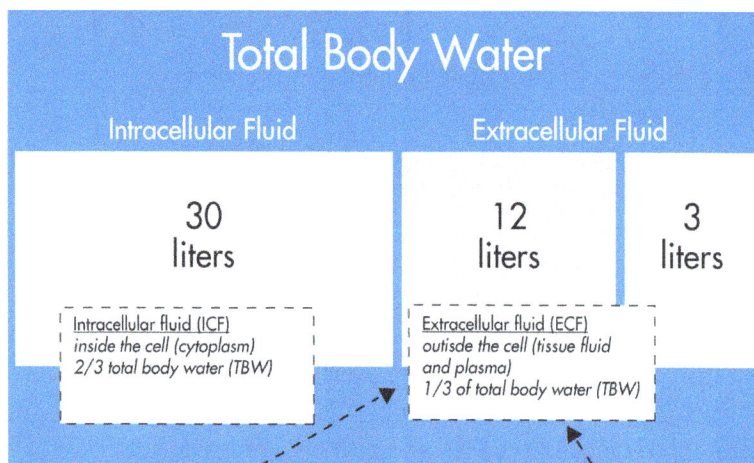

Total Body Water

Intracellular Fluid	Extracellular Fluid	
30 liters	12 liters	3 liters

Intracellular fluid (ICF)
inside the cell (cytoplasm)
2/3 total body water (TBW)

Extracellular fluid (ECF)
outisde the cell (tissue fluid
and plasma)
1/3 of total body water (TBW)

FIGURE 2.1
The human body fluid compartments.

Serology

In 1901, Karl Landsteiner made a significant discovery that led to the grouping and typing of blood. His discovery earned him the Nobel Prize about twenty-nine years later. For years physicians had attempted to transfuse blood from one individual to another and very often their efforts resulted in death of the recipient due to agglutination reactions. Landsteiner was the first to recognize that not all individual red blood cells were the same and individuals could be grouped according to specific antigens found on the red blood cell membranes. This discovery led to the ABO blood grouping system which is widely used in clinical medicine today as a means of insuring blood transfusion compatibility (figure 2.2). Eventually researchers pursued Landsteiner's work and in 1937 another major red blood cell antigen was discovered called the Rh factor (Rh from the discovery based on blood from the Rhesus monkey). Today more than one hundred different antigens (cell membrane markers) have been discovered, however the major ones in the ABO grouping system are the most important for cross matching blood between a donor and a recipient. Before the discovery of DNA fingerprinting techniques forensic scientists found that the most accurate means of identifying an individual was through the use of the ABO blood grouping system. Scientists recognized that no two people (except identical or monozygotic twins) would possess exactly the same red blood cell surface proteins (antigens). Since these blood antigens are controlled genetically they become a unique feature of each individual. This becomes extremely relevant in a forensic sense especially when violent crime scenes often involve the exchange of blood and other body fluids. Detecting the presence of blood on a victim or an assailant can become useful evidence in identifying a specific individual. The following chart (Table 2.1.) shows the antigenic relationships between the various antigens and antibodies for the four major blood groups: A, B, AB, and, O. Today as a result of DNA technology blood protein identification is not as

Table 2.1 Major antigens and antibodies in the ABO groups.

ABO Group	Antigens on RBC's	Antibodies in Serums
A	A	Anti B
B	B	Anti-A
AB	A and B	Neither Anti-A or Anti-B
0	Neither A nor B	Anti-A and Anti-B

reliable as DNA evidence so in most criminal forensic cases the ABO blood antigens are much less significant than in the past.

The genetics of the ABO blood grouping system is worth exploring in order to be able to look at paternity issues as one example. Understanding some basic human genetic concepts is essential for understanding the inheritance of specific blood groups. Issues here involve understanding a few key concepts in genetics:

1. Genes are fragments of DNA and are found linked on individual chromosomes.
2. Chromosomes and genes always occur in pairs (one inherited from each parent).
3. Genes segregate and assort independently during meiosis (sex cell formation).
4. When both genes are the same on each chromosome the individual is homozygous.
5. When both genes are different on each chromosome the individual is heterozygous.
6. Genotype is the type of genes present on the chromosomes.
7. Phenotype is the outward appearance of the trait or characteristic.

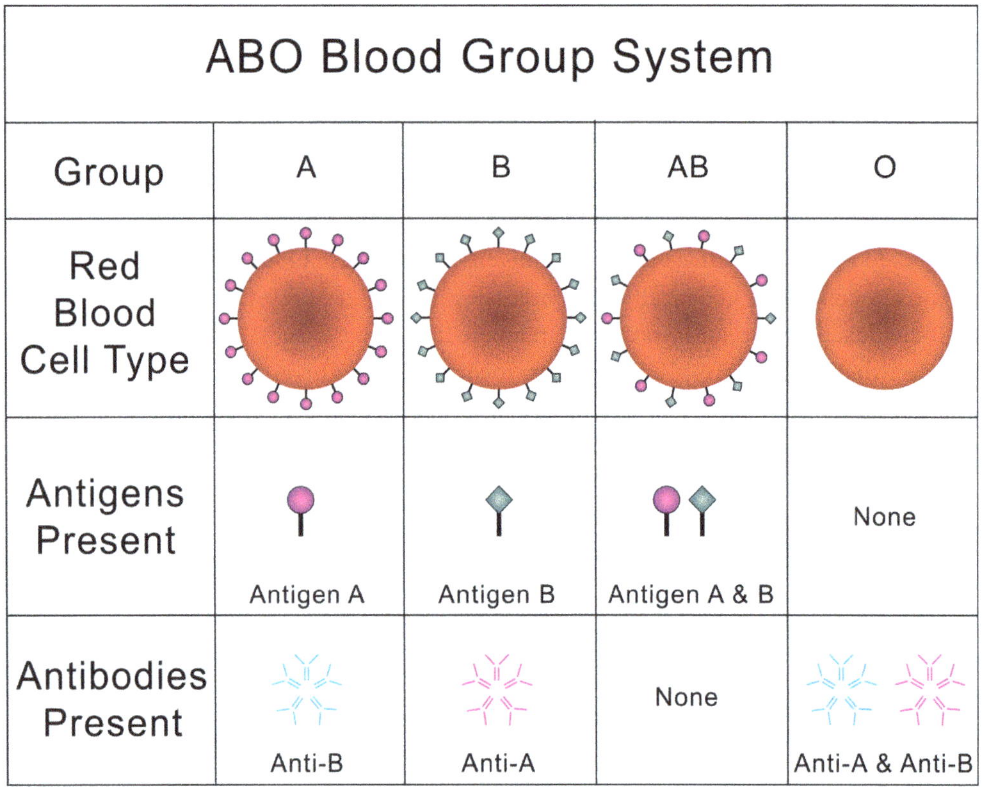

ABO Blood Group System				
Group	A	B	AB	O
Red Blood Cell Type				
Antigens Present	Antigen A	Antigen B	Antigen A & B	None
Antibodies Present	Anti-B	Anti-A	None	Anti-A & Anti-B

FIGURE 2.2
Antigen antibody reactions

Table 2.2 shows how the arrangement of the chromosomes and the individual symbols for each of the major ABO antigens. Note that by convention the capital letter I is used to carry the gene and the system is codominant so that there are two major genes which relate to the four possible ABO groups. If neither of the two genes A or B are present the individual is purely recessive and is characterized as having a genotype ii and the phenotype Group O. The genes that influence the presence of the Rh factor work in the same way as the ABO blood antigens. In addition to the presence of the surface antigens on the red cells there are also corresponding antibodies found in the blood plasma. People who are group A have antibody B in their blood plasma, people who are group B have antibody A in their plasma, people who are group AB have neither antibody in their plasma and people who are group O have both antibody A and antibody B in their plasma. Figure 2.2 shows the antigen-antibody reactions which result between the recipient and donor cells when transfusing blood.

When antibody A reacts with antigen A an agglutination reaction occurs which results in clumping of the blood cells. As seen previously in figure 2.2 antigen antibody reactions occur when the corresponding serums with antibodies are combined with the antigens on the red blood cells. Since group O blood has neither of the two surface antigens *a* or *b* it becomes the universal donor and group AB becomes the universal recipient. Remember it is the antigens on the cell membranes which are primarily responsible for binding the antibodies to cause the agglutination reaction. Even though group O has both antibody *a* and antibody *b* it may be transfused to all other blood groups because it does not have either of the antigens on its membrane surface.

Table 2. 2 **Genotype and phenotype of ABO group.**

Group	Genotype	Phenotype
A	$I^A I^A$ or $I^A i$	A
B	$I^B I^B$ or $I^B i$	B
AB	$I^A I^B$	AB
O	ii	O

Blood as a Tissue

Blood is a highly specialized type of connective tissue and is composed of a liquid fraction called the plasma and a cellular fraction referred to as the formed elements. The plasma (which comprises approximately

THE ELEMENTS OF BLOOD

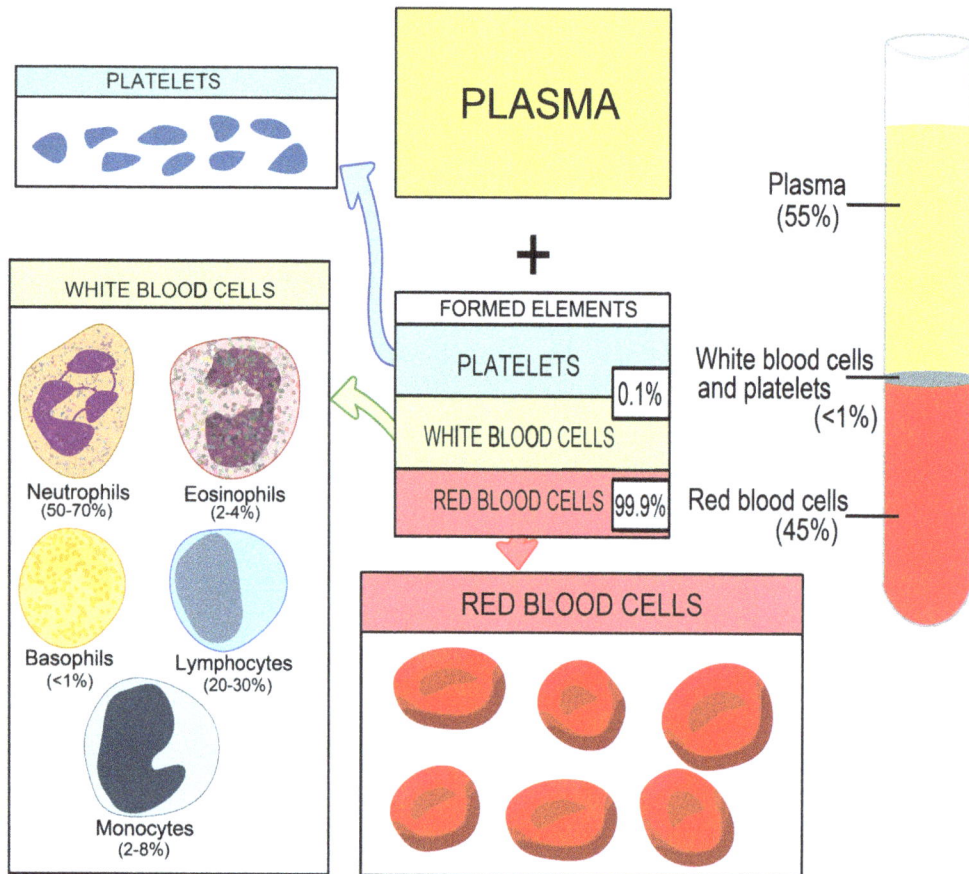

FIGURE 2.3
Plasma and its contents.

55% of the total blood volume) contains various proteins, ions, and a variety of other organic and inorganic material as seen in figure 2.3. The formed elements (which comprise approximately 45% of the cellular fraction including the erythrocytes (RBC's), leukocytes (WBC's), and thrombocytes (platelets).

The origin of the formed elements is the various components of red bone marrow found throughout the skeletal system. The hemocytoblast (stem cell) is the progenitor cell of all of the formed elements. Figure 2.4 describes the various types of white blood cells.

WHITE BLOOD CELLS

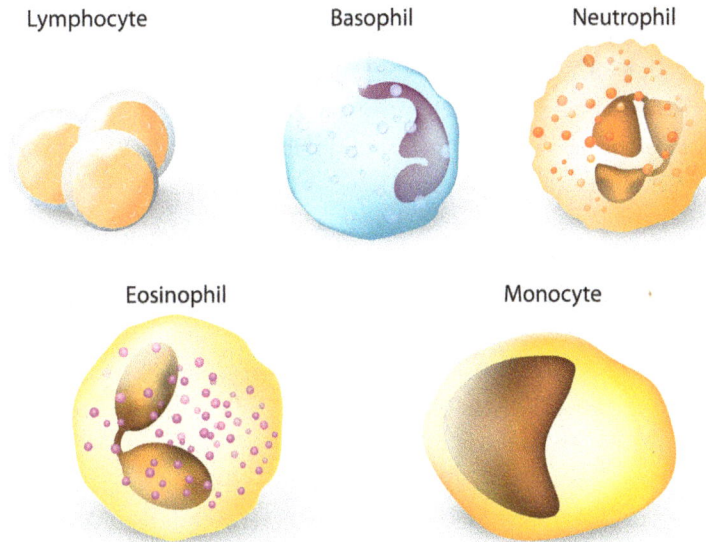

Lymphocyte   Basophil   Neutrophil

Eosinophil   Monocyte

FIGURE 2.4

Various types of White blood cells.

Differentiating the mature leukocytes (WBC's) into the granulocytic and agranulocytic series become important in making many medical diagnoses. The normal percentage of each of these cells is seen in Table 2.3 and some examples of abnormal WBC differential counts are also discussed in the following paragraph. Additionally, the normal erythrocyte (RBC) and thrombocyte (platelet) counts are included in the table, each of which has significant meaning in medical diagnosis.

Normal WBC counts, being approximately 5000-7500 cells/mm^3, are extremely significant in diagnosing many disease states. Elevated WBC counts up to 20,000-30,000 cells /mm^3 may

Table 2.3 **Normal percentages of cells.**

CELL	ERYTHROCYTE	NEUTROPHIL	EOSINOPHIL	BASOPHIL	LYMPHOCYTE	MONOCYTE
%	$4.5\text{-}5.5 \times 10^6$	60%	1-2%	0.5-1%	30%	5-10%
#	Per mm^3	3000/mm^3	5-10/mm^3	3-5/mm^3	1500/mm^3	250-500/mm^3
	Blood	Blood	Blood	Blood	Blood	Blood

indicate infection or possibly some form of leukemia. Elevated numbers of lymphocytes are often seen in certain types of viral infections, elevated eosinophils are often seen in cases of immunological responses such as allergic reactions and asthma. Platelets are small cell fragments which come from megakaryocytes that break off in the bone marrow and become part of the circulating blood. Their normal counts range between, 250,000-400,000 cells/mm³. Decreased numbers of circulating platelets may result in the inability to clot or coagulate blood and increase the risk of hemorrhage. Increased numbers of platelets may lead to intravascular clotting and/or the formation of vascular plaques which cause narrowing of the lumen of the blood vessels that in turn could lead to stroke or cardiovascular infarction.

Differentiating human somatic cells as to whether they are male or female is possible by looking for the presence of a Barr Body, named for a Canadian anatomist M.L. Barr in 1949 (Mary Lyon Hypothesis) especially in the nuclei of neutrophils. In females one of the X chromosomes is inactive and condenses and appears as a small out pocketing along the edge of the nuclear membrane. The identification of Barr Bodies is indicative of the cells having come from a female. They do not appear in the XY male.

Antigen - antibody reactions have become very important in areas of serology other than being used to determine ABO blood groups. In the forensic lab the use of specific antibodies has been used to identify drugs and other components in blood and urine. By combining a specific drug with a protein and injecting it into another animal (usually rabbit or mouse) these animals will produce specific antibodies to the known drug complex. Removing the blood serum from the animal provides a valuable source of antibodies that may be used to detect various substances in blood and urine. Enzyme-multiplied immunoassay technique (EMIT), enzyme linked immunosorbant assay (ELISA) and radioimmunoassay (RIA) are all modern used methods to detect and quantify substances in body fluids using the principles of antigen-antibody reactions.

Another issue in forensic investigations is to determine whether or not a substance is blood or some other foreign material. Additionally, if the substance is determined to be blood is it human blood or some other type of animal blood. To detect blood forensic scientists use a variety of chemical tests which are known as presumptive tests. These presumptive tests for blood rely on the fact that hemoglobin has the ability to catalyze the oxidation of certain reagents. In most cases the oxidizing agent is 3% hydrogen peroxide (H_2O_2). One common test to identify an unknown substance as blood is the Kastle-Meyer Test. In the Kastle-Meyer test a drop of a solution of phenolphthalein (colorless) is applied to the unknown stain. Next, a drop of hydrogen peroxide solution is added. The formation of a bright pink color radiating from the reaction sight is indicative of the presence of blood. Other tests such as leucomalachite green (LMG) are also used to presumptively identify the presence of blood. In some cases blood is not apparent because it has been attempted to be cleaned at the crime scene so a reagent such as Luminol may be used for detecting the presence of latent blood patterns. All of these tests are presumptive because there are a number of plant materials such as some vegetables, household bleach and a few metals that could give a false positive result. Once it has been determined that the foreign substance is blood a serologist may use specific antibody tests and DNA analysis to determine if the origin of the blood is from a human or some other animal source. The precipitin test, is a reaction between human red blood cell antigens and human antiserum (produced by injecting human blood into a rabbit) and is a means of positive identification of human blood.

Urine

Urine is a body fluid that is produced by the kidneys as a result of normal cellular metabolism. The kidneys filter blood through a vascular system that is integrated with the filtering units in the kidneys which are called nephrons. Figure 2.5 shows the location of these nephrons and their arrangement within the kidney tissue. There are approximately 1,500,000 nephrons found in each kidney. On a daily basis, considering normal metabolism and fluid compartment balance, the kidneys filter approximately 180 liters of fluid (blood plasma) and at the end of the filtration process approximately 178.5 liters are reabsorbed. This results in a normal urine output of approximately 1.5 liters which would be considered to be a normal 24 hour urine output in medical physiology. Normal urine appears straw colored, clear, pH between 4.5-6.5, specific gravity between 1.005-1.025 and is comprised largely of urea, various ions, and water.

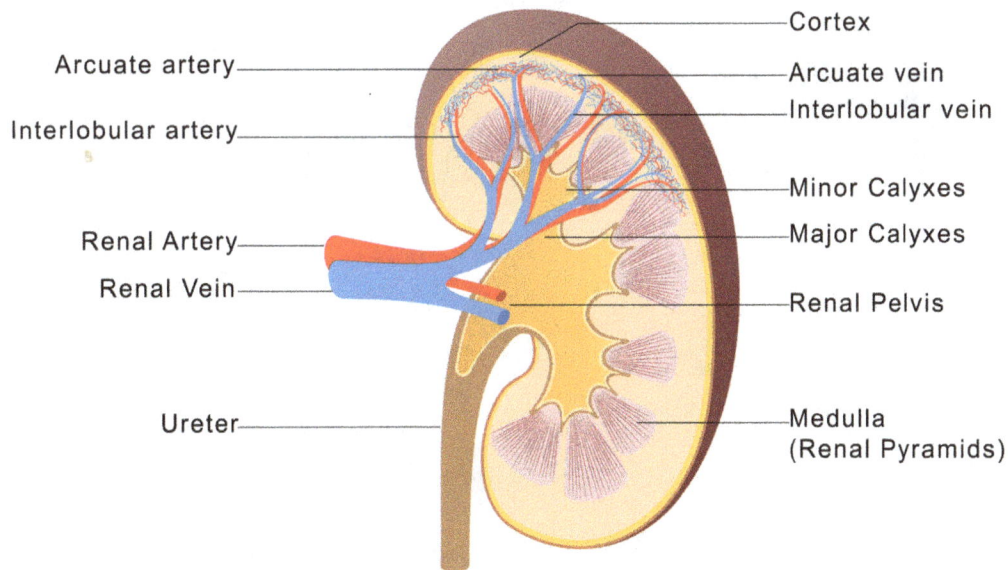

FIGURE 2.5
The kidney.

In a forensic sense the urine could be analyzed for various drug metabolites, poisons, glucose, proteins, bilirubin and elements such as crystals, casts and foreign substances in the sediment. Ethyl alcohol causes diuresis of the kidneys and produces large volumes of relatively colorless urine so upon examination of a urine sample from an individual who might be inebriated the urine appears to be colorless. This information might support further tests of blood alcohol levels and corroborate alcohol toxicity. The urine like the blood becomes a valuable body fluid for detecting various metabolic substances that might be linked to the cause of death.

Semen

Because of the large number of crimes that are of a sexual nature the identification of seminal fluid is important to the forensic scientist. Semen is a fluid produced by the testes, bulbourethral glands, seminal vesicles and the prostate gland. Semen is largely (90%) water and has a slightly alkaline oh between 7.2-7.4. One milliliter of semen may contain between $20\text{-}100 \times 10^7$ spermatozoa and a normal male ejaculates between 3-6 ml per ejaculation. Some males have a diminished sperm count which is referred to as oligospermia and in some males who produce no sperm at all the condition is referred to as azoospermia (this condition may exist in the case of a vasectomy). The presence of seminal fluid may be detected with an ultraviolet light as the

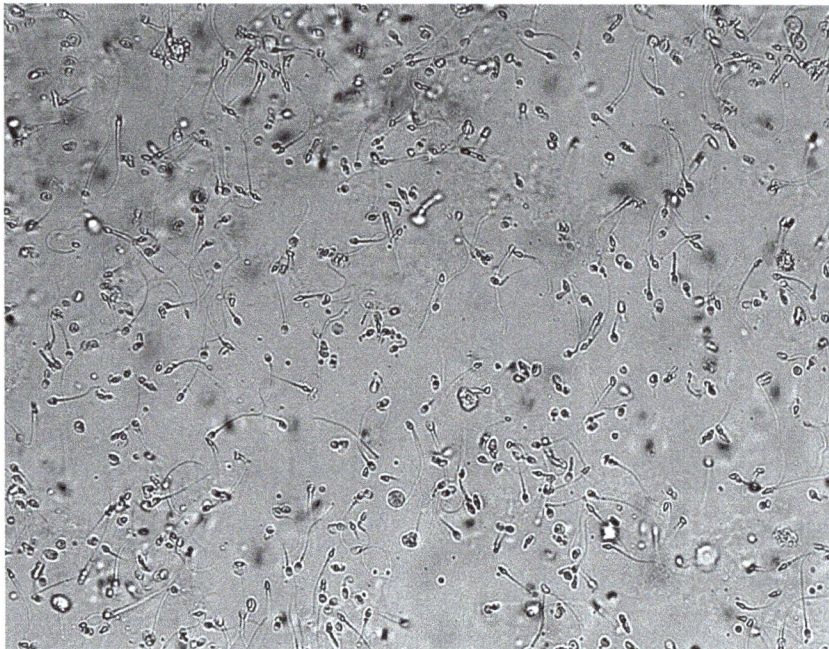

FIGURE 2.6
Sperm cells under electron microscope.

semen will fluoresce under UV light intensity. This is a presumption test since urine will also produce fluorescence under UV light. A common presumption test for identifying seminal fluid is the acid phosphatase test. Acid phosphatase is an enzyme produced by the prostate gland and is secreted in very high concentrations in comparison to any other body tissue. It is presumptive because certain fruit juices, vaginal secretions and vaginal creams may also give a positive result. Another presumption test which is used to detect semen is the PSA or the p30 test. The enzyme is produced by the prostate gland. The p30 is found exclusively in seminal fluid so vaginal stains will not give a positive test, unlike the acid phosphate color test. Perhaps the most definitive test for the presence of semen is to identify the actual spermatozoa. Sperm cells may be obtained by wetting the source and transferring the fluid to a glass slide and observed under the microscope. Sperm cells are extremely small and the flagella accounts for most of their 40-60 micrometer length as is seen in the picture in figure 2.6.

Saliva

Saliva is a fluid or secretion that is a serous mucous material that is produced by the parotid, submaxillary and sublingual salivary glands. It is comprised of about 99% water, has a pH in the range of 6.8-7.0 and in the average adult approximately 1.0-1.5 liters are secreted each day. Saliva contains the enzyme amylase which can readily be detected by using the starch-Lugol's solution test. Amylase in the presence of starch will digest the starch into sugar and if Lugol's solution is place on a starch source to which saliva has been introduced the color change will be negative. Lugol's solution turns blue/black in the presence of starch and remains reddish/brown in its absence. The significance of finding saliva, for example on the rim of a glass, an envelope, a cigarette filter or some other material where human saliva might have been deposited leads to a potential source of cheek epithelial cells which will provide DNA for testing and as a possible means of identification.

FOR FURTHER READING

Huff, E. (2012). New York court rules bodily fluids are not deadly weapons. *Natural News.* Retrieved from http://www.naturalnews.com/036221_HIV_positive_assault_bodily_ fluids. html#ixzz26MzK9MHP

1. The largest of the body's fluid compartments is the ICF, which contains approximately _____ L of fluid.
2. There are on average about _____ L of plasma fluid in the vascular fluid compartment.
3. Upon centrifugation, the three major components of whole blood are: _____, _____, and _____.
4. Red blood cells may be identified based on the types of surface _____ they display.
5. An individual who has group B blood would show the presence of _____ antigens on the red blood cells.
6. During the ABO blood grouping procedure, antibody-A and antigen-A will react and the mixture will clump or _____.
7. In a paternity case a mother presents herself as a group B and has a child who is group O. The alleged father is also group B. Is it possible that the child could be from this union?
8. The presence of a Barr body on the nucleus of a neutrophil might indicate the blood is that of a _____.
9. In studying the chemistry of the blood, the portion of interest is the _____.
10. A positive Kastle–Meyer test indicates the presence of _____.
11. The Rh factor is another antigen found on certain red blood cells. If the Rh protein is present, the individual is considered type _____.
12. The average 24-hour urine output in an adult is approximately _____ L.
13. Acid phosphatase and PSA tests are useful for the identification of _____ fluid.
14. The importance of identifying the presence of saliva at a crime scene is that we may possibly find the presence of _____ cells, which could be used for DNA analysis.
15. A normal RBC count is between 4.5 and 5.5×10^6 cells per cubic _____ of blood.

BLOOD SPATTER PATTERN ANALYSIS

LEARNING OBJECTIVES

After completion of this chapter, students will be able to:

» Explain the nature of fluids and how they behave
» Relate the methods by which blood spatter is produced
» Differentiate the spatter patterns of low, medium, and high velocity impact spatter
» Identify the height and angle of impact of a spatter pattern
» Measure the distances and angles to construct the scene of a violent crime
» Differentiate various patterns of blood spatter
» Explain the spatter pattern in relation to the surface texture
» Arrange spatter patterns based on angle of impact

The geometric interpretation of blood stain patterns has been scientifically investigated and recently it has gained a much greater recognition. A prominent blood stain analyst, Herbert Leon MacDonell (chapter 2 in Dead Reckoning) is a leading expert in the United States on blood stain pattern interpretation. He conducts an educational institute in Corning, New York where people train to become experts in the field. The use of bloodstain pattern interpretation as a recognized forensic discipline in the modern era dates back to 1955 when Dr. Paul Kirk of the University of California at Berkeley submitted an affidavit of his examination of bloodstain evidence and findings in the case of the State of Ohio v. Samuel Sheppard. This was a significant milestone in the recognition of bloodstain evidence by the American legal system.

The biological properties of blood have been covered in the previous chapter. The physical properties of exposed human blood are similar to most other fluids such as water. Exposed blood will act in a predictable manner when subjected to external forces. Blood is a fluid tissue and as such adheres to most of the laws of fluid physics. Whether a single drop or a large volume it is held together by strong cohesive molecular forces that produce a surface tension within each drop and on the external skin or surface. Surface tension is a force that tends to pull the surface molecules in a liquid toward the interior, decreasing the surface area and causing the liquid to resist penetration. The surface tension of blood is slightly less than that of water. The surface tension of liquid mercury is 10 times that of water which makes a drop of mercury very difficult to break apart (see demonstration). To create blood spatter some force must act on the volume of blood to overcome its surface tension. The typical spherical shape of a blood droplet travelling in space is directly related to the molecular cohesive forces acting upon the surface of the drop. One false assumption regarding free falling fluids is that they appear tear dropped in shaped and this is not true as all fluids take on the spheroid shape.

Basic principle #1 is that a free falling drop of blood forms a sphere or a ball. A passive drop of blood in space is created when the gravitational forces acting on the blood overcome the molecular cohesive forces and the drop breaks away from its source. The pictures above (figure 3.1) shows a time lapsed blood droplet impacting a surface and producing castoff droplets which will result in the formation of satellite droplets. A drop of blood falling from a finger will be larger than one originating from a hypodermic needle and will be smaller than a drop coming from a baseball bat (figure 3.2).

The mutual attraction of the molecules of blood (viscosity) is defined as resistance to the form or flow. The more viscous a fluid the more slowly it flows. Blood is approximately 5-6 times more viscous than water and has a specific gravity slightly higher than water. Specific gravity is defined as the weight of a substance relative to the weight of an equal volume of water. These physical properties of blood tend to maintain the stability of exposed blood or blood drops and cause them to resist alteration or disruption.

A blood drop falling through air will increase its velocity until the force of air resistance that opposes the drop is equal to the force of the downward gravitational pull. To this point, the drop

FIGURE 3.1
Drop of blood.

achieves its terminal velocity. In his early research MacDonell established that the maximum terminal velocity for an average-sized (0.05 milliliter) free falling drop of blood was approximately 25.1 feet per second and is achieved in a maximum falling distance of between 2—25 feet. The resulting diameter of the blood stain produced by the free falling drop of blood is a function of the volume of the drop, the surface texture it impacts, and up to a point, the distance fallen. One can easily demonstrate that free falling drops of blood with a typical volume of 0.05 milliliter will produce bloodstains of increasing diameters when allowed to drop from increasing increments of height onto smooth, hard cardboard. The measured diameters range from 13.0 to 21.5 millimeters over a dropping range of 6 inches to seven feet. Blood drops that fall distances greater than 7 feet will not produce stains with any appreciable increases in diameter. It is not possible to establish with a high degree of accuracy the distance that a passive drop of blood has fallen at a crime scene since the volume of the original drop is no known.

Basic principle #2 states that a drop of blood will not break apart unless one of two things happens: a.) it strikes another object or surface which equates to the effect of the target surface which the drop of blood strikes; b.) it is acted upon by some force which equates to the types of impacts which are involved in a violent crime (this force creates spatter). One factor in breaking the surface tension of a blood drop is the physical nature of the target surface the drop strikes. Generally, a hard, smooth, nonporous surface such as clean glass or smooth tile will create little if any spatter in contrast to a surface with a rough texture such as wood or concrete that

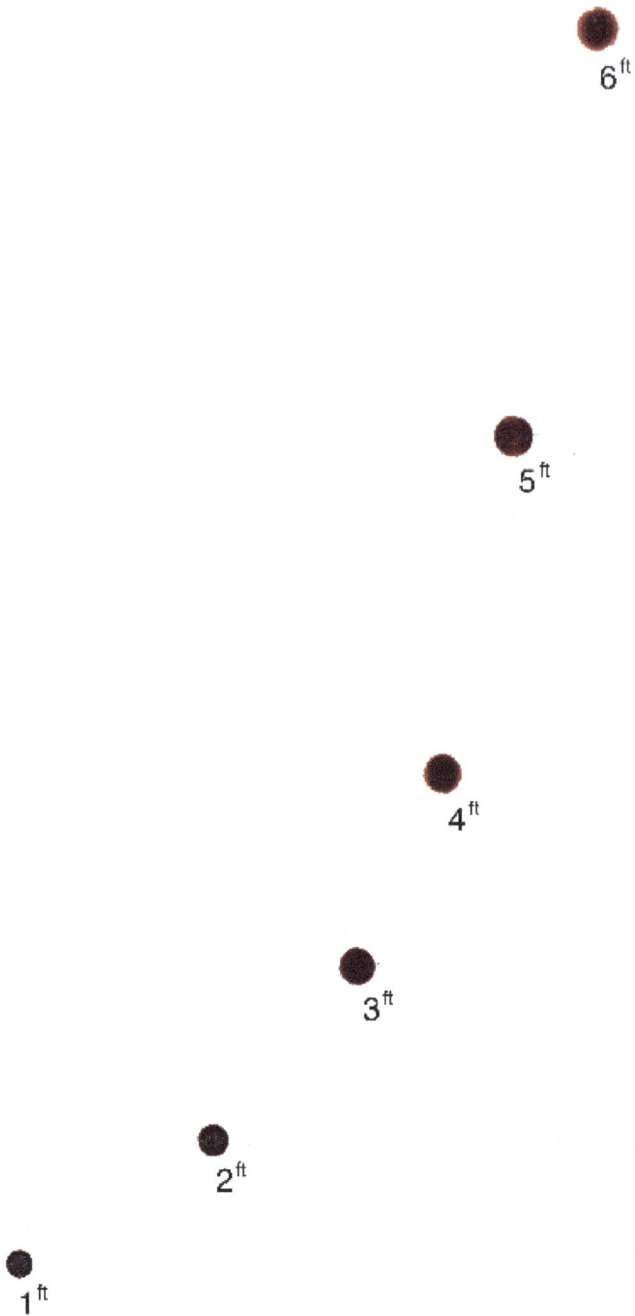

FIGURE 3.3
Drops of blood falling from a finger showing increased diameter with increased height above surface.

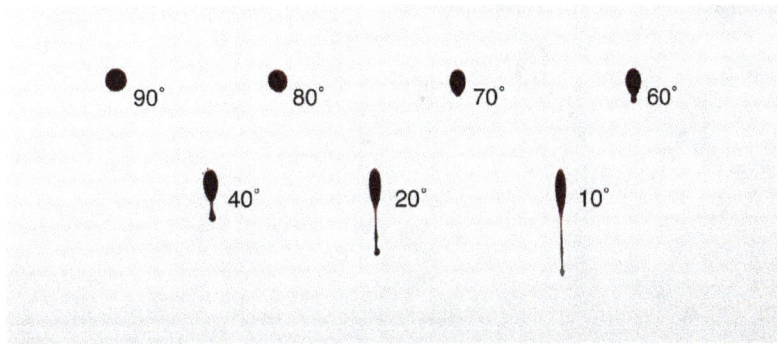

FIGURE 3.3
Drops of blood impacting a surface at various angles.

can create a significant amount of spatter. Rough surfaces have protuberances that rupture the surface tension of the blood drop and produce spatter and irregularly shaped parent stains with spiny or serrated edges.

Each of the stains in the pictures below (figure 3.3 and 3.4) were produced by drops which were identical in volume and were allowed to fall the same distance before impacting the target at specified angles.

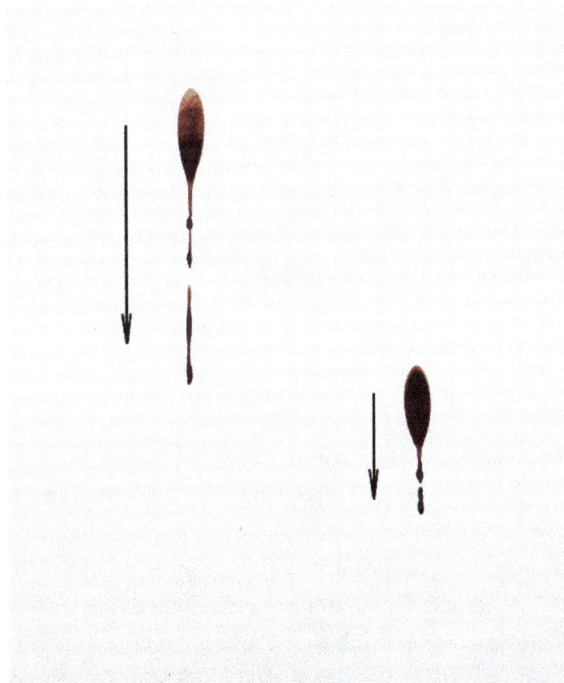

FIGURE 3.4
Angle and directionality of blood stains.

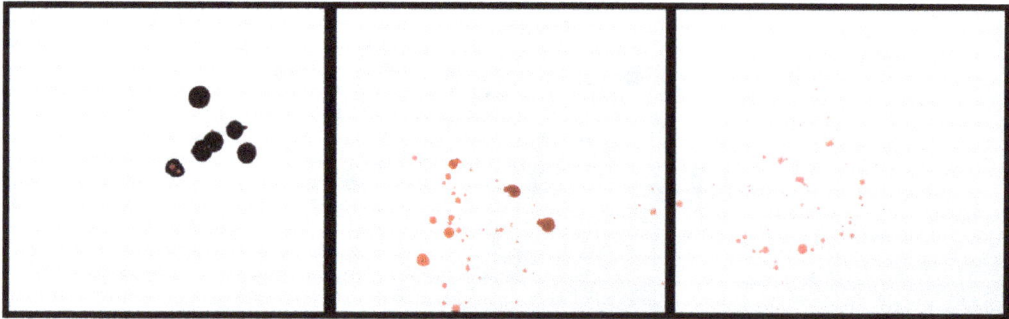

FIGURE 3.5
Low, Medium and High Velocity Spatter.

Above two drops of blood impacting a surface at a distance of 30 inches the top falling onto a smooth polished tile and the other onto wood paneling.

The size shape and directionality of blood stains are all attributable to size and velocity of the blood in flight (figure 3.4). Blood stains that are fairly regular and appear as circular patterns usually have impacted a surface at 90 degrees. As the angle of impact gets steeper such as 10 degrees the pattern of the impact spatter tends to form a leading edge which is more tapered. This effect can be seen in the diagram on figure 3.4 and 3.7 which describes the angle of impact.

Spatter originates from static blood which has been acted upon by some impacting force. Note that it is "spatter and NOT "splatter". Spatter size and the velocity of the impacting force will determine the final spatter pattern on the ground, wall or ceiling. Velocity of the force applied is inversely related to the size of the spatter produced. The small force (slower moving) results in a larger spatter size and a greater force (faster moving) results in smaller spatter sized patterns. This principle allows for examination of the spatter produced and extrapolating back to the type of force which was applied to cause its formation as seen in figure 3.5.

Spatters are categorized according to the velocity of the impacting force causing their production and recorded as:
- Low Velocity Impact Spatter (LVIS)
- Medium Velocity Impact Spatter (MVIS)
- High Velocity Impact Spatter (HVIS) (Fig. 3.5 Above)

Categorization is based on the size of the majority of the spatters because it is possible that any force can produce spatters larger or smaller than what is considered "characteristic spatter size". The diagram in figure 3.5 relates the velocity of the force applied and the general spatter size.

The **point or area of origin** or the location of the source of blood in a three-dimensional perspective may also be determined. By establishing the impact angles of representative bloodstains and projecting their trajectories back to a common axis and extending at 90 degrees up from the two-dimensional area of convergence along the Z axis, an approximate location of where the source was when it was impacted may be established. Diagrammatic representations of convergence and origin utilizing the X, Y, and Z axes are shown in figure 3.6. The figure

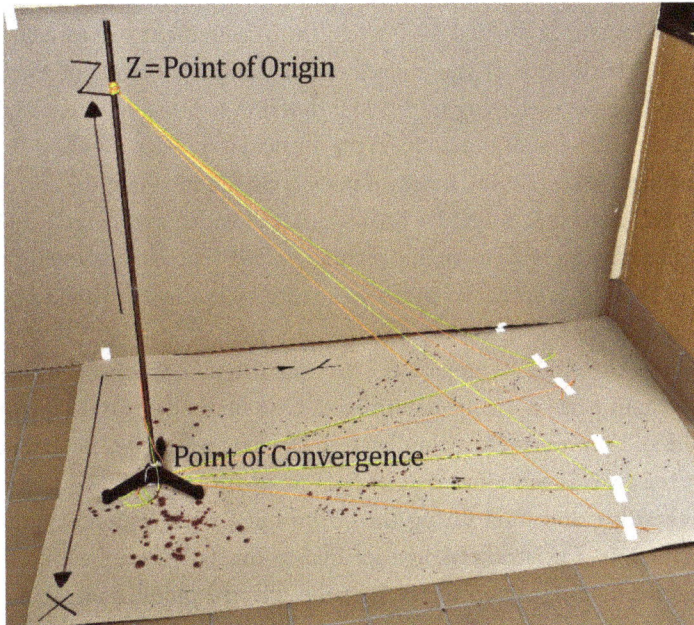

FIGURE 3.6
Points of Origin and Convergence.

identifies the point of convergence from the projected spatter patterns and the point of origin is identified along the X, Y, and Z axes. A mathematical relationship exists between the width and length of an elliptical bloodstain that allows for the calculation of the angle of impact for the original spherical drop of blood. This calculation is accomplished by measurement of the width and the length of the bloodstain as seen in the diagram on figure 3.7.

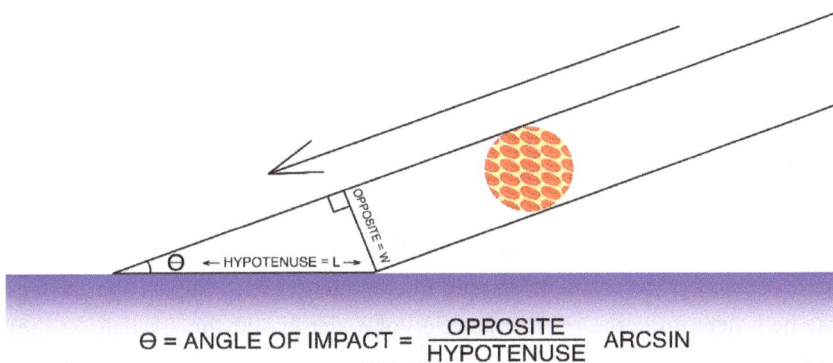

$$\Theta = \text{ANGLE OF IMPACT} = \frac{\text{OPPOSITE}}{\text{HYPOTENUSE}} \ \text{ARCSIN}$$

FIGURE 3.7
Calculation of the angle of impact.

The width measurement is divided by the length measurement to produce a ratio number less than 1. This ratio is the sine of the impact angle. The impact angle of the bloodstain may now be determined by either referring to the sine function in a trigonometric table or by using a scientific calculator with a sine function. With a calculator, after dividing the width by the length utilize the function key, arc sin, sin -1, or inverse sin function and the corresponding angle of impact will be displayed. For a circular bloodstain, the width and length are equal and thus the ratio is 1.0, which corresponds to an impact angle of 90 degrees. For an elliptical bloodstain whose width is one half its length, the width-to-length ration is 0.5, which corresponds to an impact angle of 30 degrees.

After establishing the angle of impact for each of the bloodstains, the three-dimensional origin of the bloodstain pattern can be determined. One method is to place strings at the base of each bloodstain and project these string back to the axis which has be extended 90 degrees up or away from the two-dimensional area of convergence. This is accomplished by placing a protractor on each string and then lifting the string until it corresponds with the previously determined impact angle. The string is then secured to the axis placed at the two-dimensional area of convergence. This is repeated for each of the selected bloodstains. Remember that this calculated area of origin is always higher than the actual origin of the bloodstains because of the gravitational attraction affecting the spatters while in flight. This gives the analyst the maximum possible height of the blood source. In practical terms the analyst is attempting to determine whether a victim was standing, lying down, or sitting in a chair when the blood was spattered.

Castoff bloodstain (figure 3.8) patterns are produced during a beating with a blunt object, blood does not immediately accumulate at the impact site with the first blow. As a result, no blood is available to be spattered or cast from the first blow. Spatter and castoff patterns are created with subsequent blows to the same general area where a wound has occurred and blood has accumulated. Blood will adherer in varying quantities to the object that produces the injuries. A centrifugal force is generated as an assailant swings the bloodied object. If the centrifugal force generated by swinging the weapon is great enough to overcome the adhesive force that holds the blood to the object, blood will be released from the object and form a castoff bloodstain pattern.

The blood that is released (castoff) will strike object and surfaces such as adjacent walls and ceilings in the near vicinity, at the same angle from which it is released or cast. The size, distribution, and quantity of these castoff bloodstains vary. Castoff bloodstain patterns may appear linear in distribution, and the individual stains are frequently larger in size than impact blood spatters. Castoff patterns are often seen in conjunction with impact spatters and a study of each may help determine the relative position of a victim and the assailant at the time the injuries were inflicted. Castoff bloodstains are not always present at scenes where blunt or sharp force injuries have occurred. The arc of the back or side swing may be minimal, especially in the case of a heavy blunt object. Occasionally, analysis will attempt to determine whether the

THE MECHANICS OF CAST-OFF AND PATTERN PRODUCED

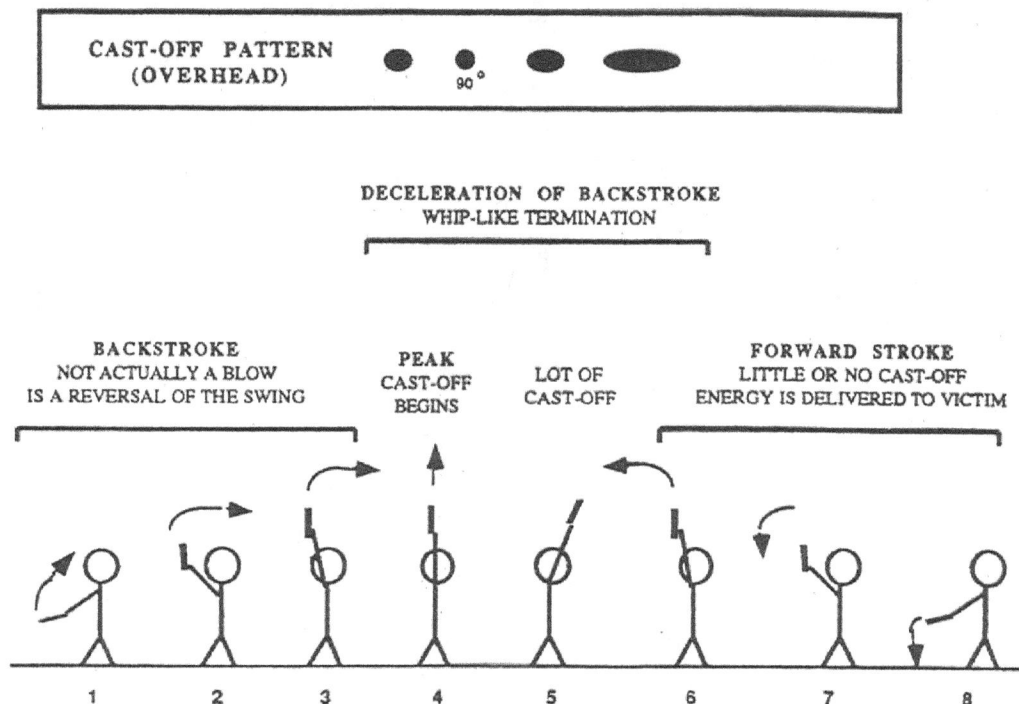

CAST-OFF PATTERN
(OVERHEAD)

90°

DECELERATION OF BACKSTROKE
WHIP-LIKE TERMINATION

BACKSTROKE
NOT ACTUALLY A BLOW
IS A REVERSAL OF THE SWING

PEAK
CAST-OFF
BEGINS

LOT OF
CAST-OFF

FORWARD STROKE
LITTLE OR NO CAST-OFF
ENERGY IS DELIVERED TO VICTIM

1   2   3   4   5   6   7   8

FIGURE 3.8
Cast off Blood Stain Patterns.

person swinging the object was right or left handed. This is dangerous since many individuals may swing object effectively with either hand. The analyst must also consider the possibility of back-handed swings that may appear similar. Other types of spatter patterns may be from expired blood which is blood that has been collected in the airways and is released upon expiration or coughing. Very often this pattern is seen to have small air bubbles within the array of small spatters. Arterial gushing is seen when an artery is severed and the blood typically takes on the pattern as seen in the picture below. This pattern is consistent with the systolic and diastolic beats of the heart showing the high pressure during systole and the reduced pressure during diastole.

Transfer bloodstain pattern occur when an object wet with blood comes into contact with an unstained object or secondary surface, a blood transfer pattern occurs. These patterns may assist an examiner in determining the object that made the pattern, i.e. hait, knife, shoe, etc., since a recognizable mirror image of the original surface or a portion of that surface may be produced.

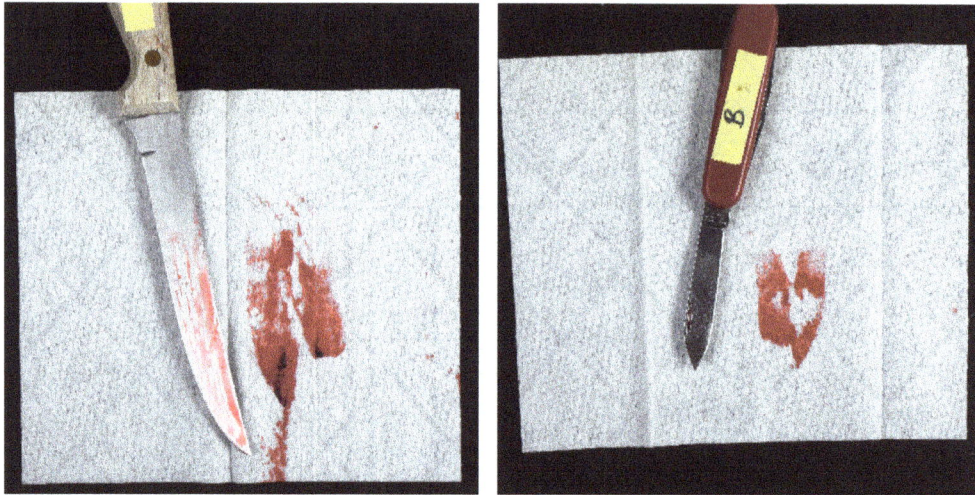

FIGURE 3.9
Bloodstain deposited on surface know as transfer pattern made by knife (swipe and wipe).

Figure 3.9 shows transfer patterns, sometimes called a swipe, of two knives that were wiped with a cloth material.

Bloodstains deposited on surfaces at a scene are subject to various forms of change from their original appearance at the time the bloodshed occurred. Recognition of these alterations and an understanding of their significance are important for the reconstruction of the event. When blood exits the body, the process of drying and clotting (coagulation) are initiated. The drying time of blood is a function of its volume, the nature of the target surface texture and the environmental conditions. Small spatters and light transfers of blood will dry within a few minutes under normal conditions of temperature, humidity, and air currents. Larger volumes of blood may take considerable time to completely dry. Drying is accelerated by increased temperature, low humidity, and increased airflow. Initially, the outer rim or perimeter of the bloodstain will show evidence of drying which then proceeds toward the central portion of the stained area.

When the center of a dried bloodstain flakes away and leaves a visible outer rim, the result is referred to as a skeletonized stain. Another type of skeletonized bloodstain occurs when the central area of a partially dry bloodstain is altered by contact or a wiping motion that leaves the periphery intact (figure 3.10). This may be interpreted as movement or activity by the victim or assailant when or after injuries were inflicted. The pattern seen below was produced by wiping a partially dried bloodstain indicating activity shortly after the blood was deposited. Notice the remaining peripheral ring of the original bloodstain caused by drying around the edges. As dried bloodstains age they tend to progress through a series of color changes from red to reddish brown and eventually to black. The estimation of the age of bloodstains based on color is

FIGURE 3.10
Sample of skeletonized blood stain.

difficult because environmental conditions in the presence of bacteria and other microorganisms will affect the sequence and duration of the color changes.

The clotting process is also initiated when blood exits the body and is exposed to a foreign surface. The appearance and extent of clotted blood at a crime scene may provide an indication of the amount of time elapsed since the injury occurred. Normal clotting time of blood that has exited the body ranges from 3-8 minutes (normal). As a clot progressively forms a gelatinous like mass it retracts and forces the serum out of and away from the progressively stabilizing clot. Any estimate of time lapse should involve experiments utilizing freshly drawn human blood of similar volume placed onto a similar surface with environmental conditions duplicated as closely as possible

The picture on the following page shows the formation of a clot after a sample of whole blood was mixed with a stick until the fibrin formation appears on the end of the stick bridging the elevated stick with the source of blood. Experimental observations of clotting processes should be compared to observations made at the scene. Conclusions should be very conservative. Evidence of coughing or exhalation of clotted blood by a victim may be associated with post injury survival time.

Existing wet blood stains at a scene are also subject to alteration in appearance due to smudging, smearing, and wiping activities of the victim or assailant. Changes in the appearance of bloodstains and patterns and additional bloodstains may also be created by paramedical treatment of the victim or removal of the victim form the scene.

Another form of bloodstain alteration is the effect of moisture, such as rain or snow, a scene exposed to the outside environment that will dilute existing bloodstains. The scene such as a vehicle may have been cleaned with water and detergents or painted after a bloodshed event. Diluted bloodstains may be difficult or impossible to evaluate without the use of a chemical enhancement process, such as luminal treatment. Also heat, fire or smoke may cause problems

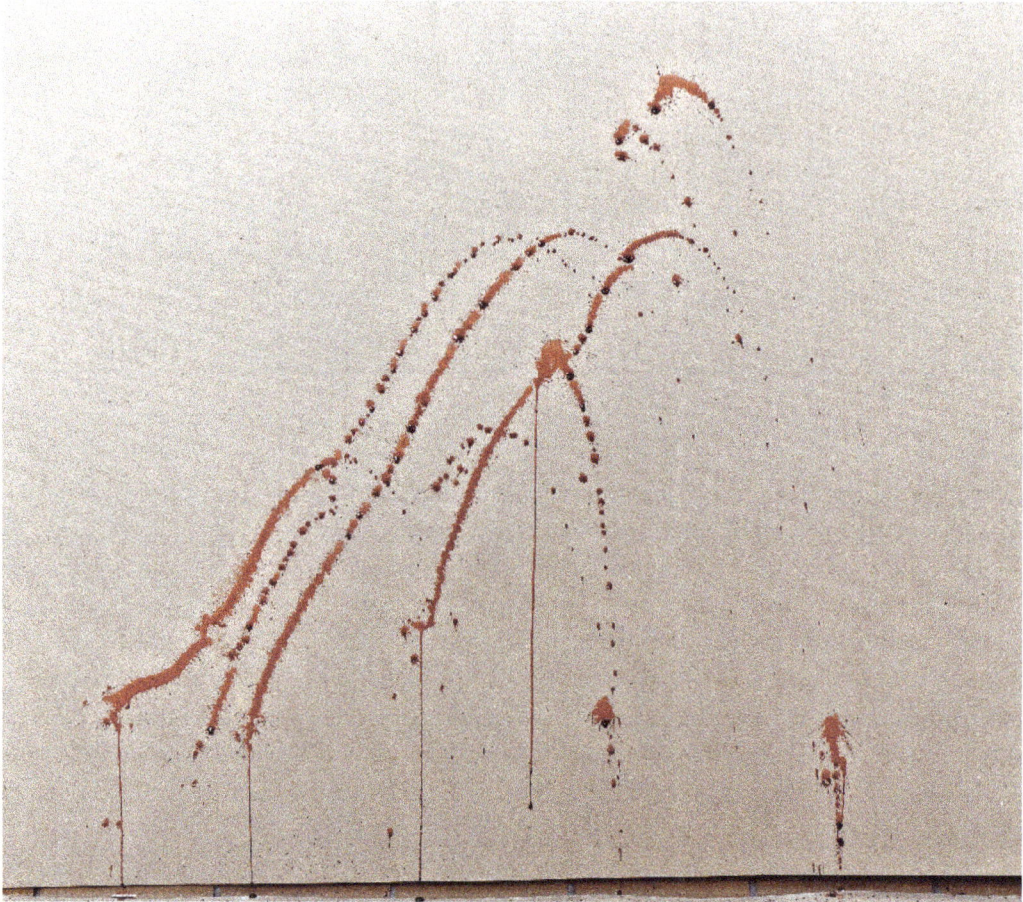

FIGURE 3.11
Arterial Gushing.

of interpretation and bloodstains covered with soot may be entirely overlooked. Heat and fire may cause existing bloodstains to fade, darken or be completely destroyed.

Blood which is projected from a ruptured or severed artery produces a pattern called arterial gushing. In figure 3.11 below the peaks and valleys of the systolic and diastolic pressure can be visualized as arterial bleeding is projected on a surface such as a wall.

1. The protein called _____ forms after a cascade of chemical events, which leads to the formation of a blood _____.

2. spatter than a surface texture that is rough (such as concrete).

3. Directionality of a blood spatter pattern may be determined by observation of the leading edge of the spatter patterns. The more elongated the leading edge indicates the _____ of travel.

4. Patterns of blood spatter produced by bludgeoning someone with a hammer are called _____ patterns.

5. Being able to calculate the angle of impact of blood spatter patterns may help in determining the _____ of the blood source as well as the point of convergence.

6. Normal clotting time for human blood is from _____ to _____ minutes.

7. Swipe patterns are produced by _____ an object such as a knife blade with a towel or rag.

8. Passive bleeding such as blood generated from a bloody nose and impacting the floor beneath would produce _____ velocity impact spatter.

9. Arterial bleeding from a wound produces a pattern called arterial _____.

10. Normal clotting time for human blood is between _____ to _____ minutes.

11. Swipe patterns are produced by _____ an object wuch as a knife blade with a towel or rag.

12. Passive bleeding such as blood generated from a bloody nose and impacting the floor beneath would produce _____ velocity impact spat

13. Wiping a dried drop of blood from a surface produces a _____ pattern.

14. Arterial bleeding from a wound produces a pattern called arterial _____.

15. The protein called _____ forms after a cascade of chemical events, which leads to the formation of a blood _____.

FORENSIC ANTHROPOLOGY

LEARNING OBJECTIVES

After completion of this chapter, students will be able to:

- » Identify the microscopic details of compact bone
- » Relate the major responsibilities of the forensic anthropologist
- » Explain the developmental hierarchy in the living world
- » Identify the major types of connective tissue and their function
- » Choose the correct anatomical term to describe location and direction on the skeletal structure
- » Differentiate male from female skeletal elements
- » Compare skeletal features among the human races
- » Name and identify the major bones in the axial and appendicular skeletons

Anthropology is technically the study of mankind. As a science it has a number of branches to include physical, cultural, archaeological, linguistics, paleontology, and a relatively new area of study called forensic anthropology which is actually a subfield in the branch of physical anthropology. The area of forensic anthropology offers a very unique humanitarian service in today's world that is over burdened with violence. The primary service of a forensic anthropologist is the description and identification of human remains that are usually beyond recognition using outward physical appearance. Analysis of human remains and remnants of artifacts that surround them may reveal considerable information which will lead to determining the time, manner and mode of death. The success of such analysis is largely dependent on the careful retrieval of the remains very often in areas where conditions are volatile and not particularly conducive for these operations.

Clandestine deaths cast a shadow on everyone. Missing persons and unidentified dead, those who have simply disappeared, are prime indicators of the worst criminal and political behavior of man. Peace and the dignity of humanity start with the efforts of forensic anthropologists to properly identify the dead and determine their fate. The make-up of this part of the population may be the bodies of derelicts that simply wandered off and died. Some of the subset may be those people who committed suicide. But many are evidence of the worst types of crime plaguing our society. They are teenagers executed by their companions; women raped my men in uniform, and possibly children abused by their caretakers. The unidentified are the evidence of serial killers who walk the streets with us and eat at the next table. In many countries the people are the ones who "disappeared" and maybe evidence of genocide and extreme abuse of political power. The one thing that binds a common thread between all of the dead is its silence. Although it may seem that all dead bodies are silent an unidentified body remains even more silent. Families of missing persons often say that they experience a pure sense of relief when the bodies of loved ones are located and identified. This brings a sense of closure and possibly empowerment through the process of funeral rituals.

Forensic anthropology is the scientific discipline that uses the methods of a physical anthropologist and an archaeologist to the collection and analysis of information about skeletal remains that may be useful in a medical-legal environment. The main objectives of an anthropological investigation is to provide information for the purposes of identification of the individual, estimation of the time, manner and mode of death, and collection of any other physical material that might be of value in reconstructing the series of events that led to the individuals death. Questions that are often answered are those such as:

- Are the remains human
- How many individuals are present among the remains
- What did the person look like, sex, race, age, height, physique, handedness
- Who is the individual
- When did the death occur

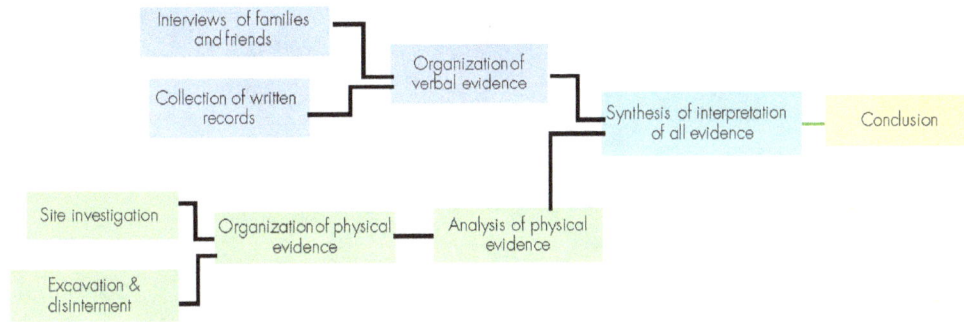

FIGURE 4.1

Flow Chart delineating orderly stages of forensic investigations.

- Did the person die at the place of burial or was he/she transported after death
- What was the cause of death (i.e., homicide, suicide, accident or natural)

Forensic anthropological investigations usually involve three major stages. Important procedures would be to collect all the verbal evidence from witnesses in the area. The next two stages would be to collect all the available physical evidence and perform an analysis of this physical evidence.

The flow chart in figure 4.1 delineates the orderly stage of a forensic investigation culminating in a synthesis and interpretation of the available information.

Considering the final stages of decomposition of the human body to be those of the remaining skeletal elements the forensic anthropologist must be thoroughly familiar with all of the details of the human skeleton considering both microscopic and macroscopic views. **Osteology** is the basic science and study of bones. It is the science that studies the development, structure, function and the variation of bone structure among individual skeletons. A complete understanding of the human skeleton is certainly a prerequisite for interpreting the history of man. Because it is not a soft tissue and it resists morphological breakdown it is often the only remaining record of man's life on earth.

Depending on the available ante mortem information the forensic anthropologist; may be able to provide the following information from examining skeletal remains:

- A description of the living person
- An evaluation of the health of the deceased
- Recognition of individuals' activities
- Identification of the deceased person
- Recognize the cause, manner and mode of death
- Determine the approximate time since death and postmortem events that occurred

There is a hierarchy of development in the human body which begins at the molecular level and advances to the complete organism. This hierarchy is seen in figure 4.2; to comprehend the structure and function of the entire organism we must be familiar with the levels of biological organization which account for its functioning as a total organism.

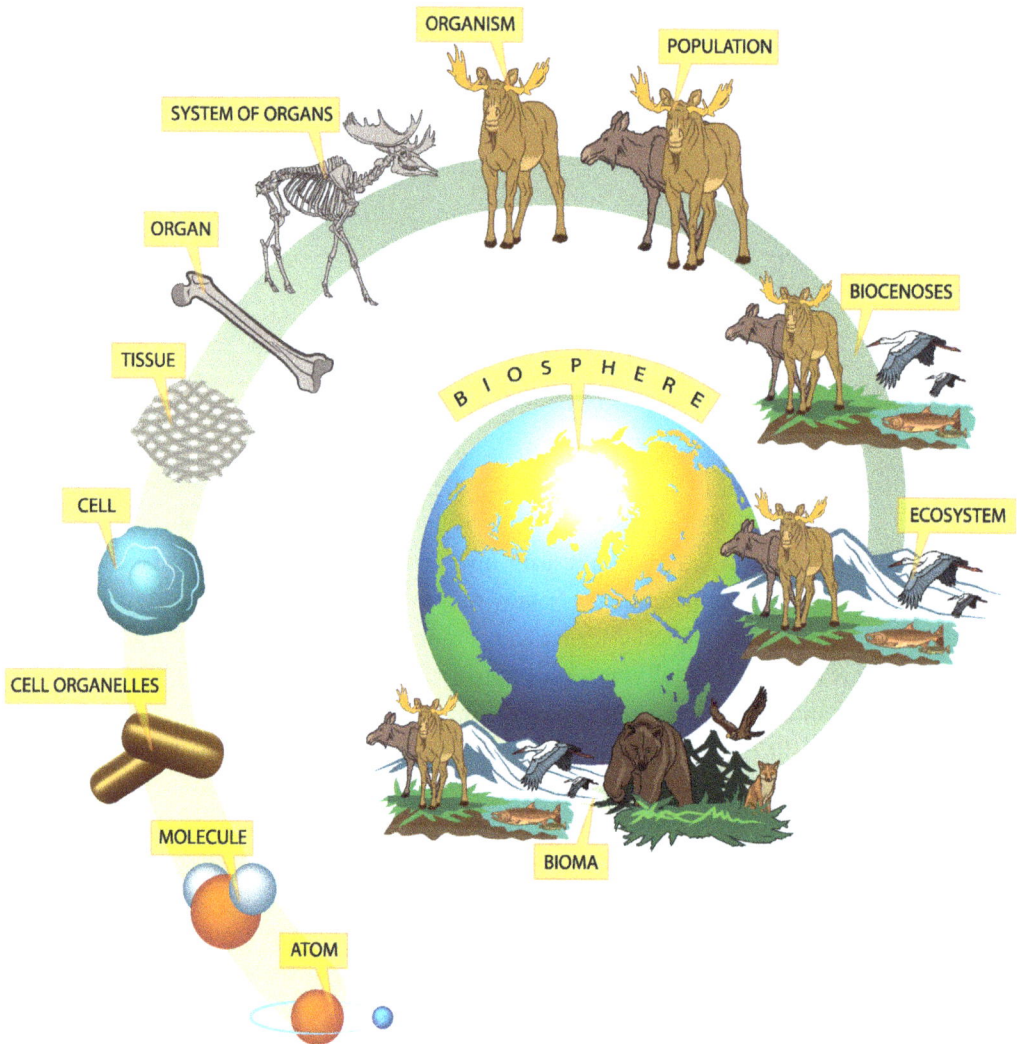

FIGURE 4.2
Biological organization.

We will begin our description of the skeletal elements by starting at the tissue level from the diagram above. A tissue by definition is a group of structurally and functionally related cells. The cells are bound together in an extracellular matrix composed of non-living materials and the various matrices have considerable chemical differences. The body's organs are built from tissues and most organs contain all four tissues which can be seen in table 4.1. It is interesting to note that as complex as the human body is, there are only four different types of tissues; however, within each group there is a high degree of differentiation and specialization.

Table 4.1 **Basic tissue types.**

Basic Tissue Types	Tissue Functions	Examples
Epithelial tissue	Covering	Skin, hair, nails
Connective tissue	Support, protection	Bone, cartilage, fat, blood
Muscle tissue	Movement	Muscle
Nervous tissue	Control	nerve

The two major elements of the skeletal system are bone and cartilage which are both highly specialized forms of connective tissue. Connective tissue proper provides a felt like network which supports many of the soft organs of the body.

Further classification denotes connective tissue to be either loose or dense. Loose connective tissues are areolar, adipose, and reticular. Dense connective tissues contain more collagenous fibers than those of loose connective tissues. The relatively thick collagen fibers can resist very strong shearing and pulling forces. Dense connective tissues may be classified as irregular such as that found predominantly in the dermis of the skin or within organ capsules. Connective tissue may also be classified as regular such as that found in ligaments and tendons or elastic which is found in some ligaments connecting adjacent vertebrae. Ligaments attach or connect bone to bone, cartilage and other structures. Tendons attach or connect muscle to bone and they appear to be narrower than ligaments and more cordlike. The predominant cell type found in all connective tissues is the fibroblast (when mature or quiescent they are called fibrocytes) which synthesizes and secretes the collagen used to build the tough connective tissue fibers. The pictures in figure 4.3 show varieties of connective tissue.

Cartilage consists primarily of water by weight (60%–80%) and is produced by cells that are called chondroblasts. These cells produce and secrete the matrix material for the pliable cartilage which is a complex material composed largely of chondroitin sulfate. Cartilage is quite resistant to tension because of the imbedded collagen fibers. It is not particularly good at resisting shearing forces (twisting and bending) which accounts for the high incidence of torn cartilages in sporting events. Cartilage is avascular and has a rather poor ability to repair itself after injury. Cartilage provides a cushioning at joints in the skeletal system and also provides a framework for structures that protrude from the body such as the nose and ears. Cartilage may be further classified as hyaline or articular cartilage is found covering the ends of long bones, building the lower structure of the nose, connecting some of the ribs to the sternum and a large quantity is found in the developing fetal skeleton. Elastic cartilage is similar to hyaline however it is imbedded with a large concentration of elastic fibers consisting of the protein elastin. Elastic cartilage is typically found in the epiglottis and the outer ear (pinna). A third type of cartilage called fibrocartilage has an array of dense collagen fibers embedded in its matrix. It forms the intervertebral discs, pubic symphysis and some tendon insertions. Cartilage cells may be one of three varieties depending on its function: chondroblasts which are cartilage building cells,

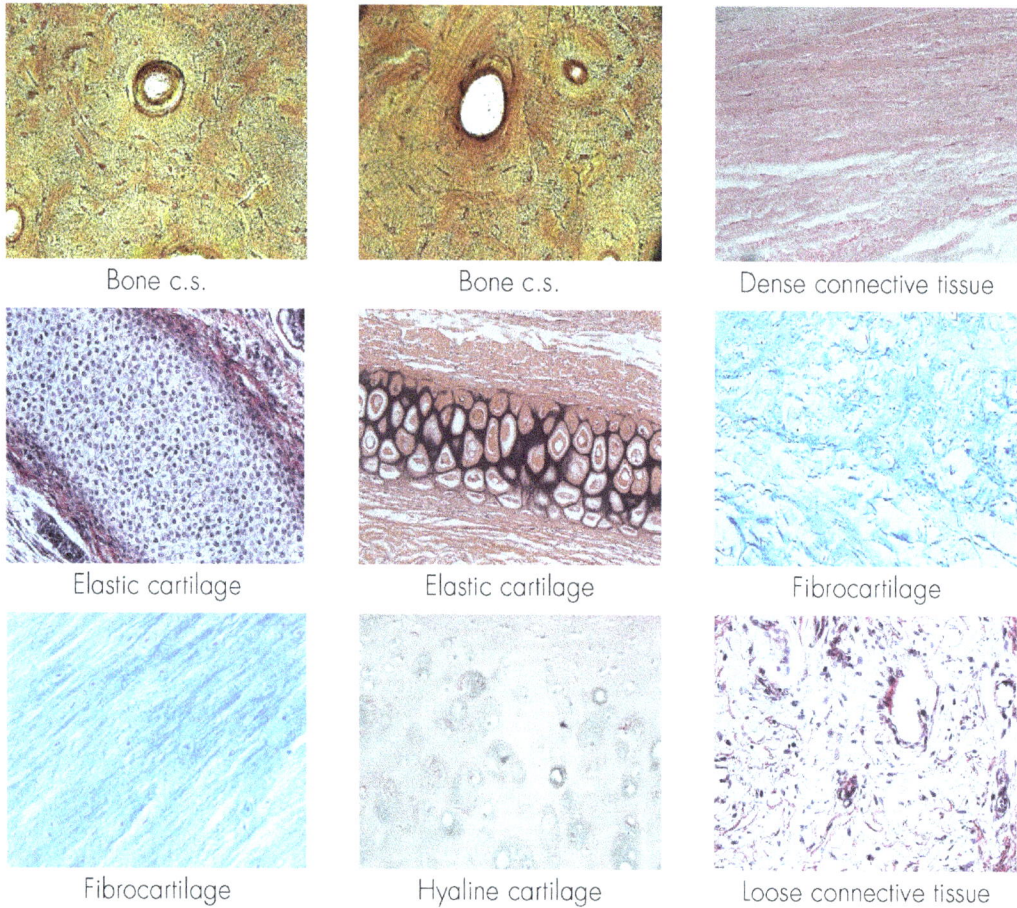

Bone c.s.

Bone c.s.

Dense connective tissue

Elastic cartilage

Elastic cartilage

Fibrocartilage

Fibrocartilage

Hyaline cartilage

Loose connective tissue

FIGURE 4.3
Various types of tissue.

chondroclasts which are cartilage remodeling cells or chondrocytes which are mature cells help-ing to maintain the normal homeostatic function of cartilage tissue. The pictures in figure 4.3 show hyaline, elastic and fibrocartilage.

The primary function of bone is for support and to provide articulations which allow for spe-cific movements through the use of various lever systems and skeletal muscle. Protection is also provided by arrangements of bones such as the skull, rib cage and pelvic girdle. A microscopic section of compact bone reveals the cellular relationship with associated matrix. Bone is a very dense tissue containing the connective tissue protein called collagen and largely inorganic salts in the form of calcium and phosphates. The complex nature of the extracellular matrix of bone consists of a tricalcium-phosphate material known as hydroxyapetite. Next to tooth enamel, bone is the hardest material in the human body. Mature compact bone develops from cartilage when blood vessels invade specific sites and begin the conversion of cartilage to bone. Bone cells that build new bone are called osteoblasts and when they mature and become permanently encircled with bony matrix they sit in small pits called lacuna and are referred to as osteocytes. During bone remodeling some of the bone cells are called osteoclasts and they have the ability to break down bone so that the osteoblasts can reform and reshape the structure. Figure 4.3 shows a microscopic section of compact bone with the typical units of the bony matrix called Haversian systems.

The long bones in the body are formed by a process called endochondral ossification whereas some of the flat and irregular bones are formed by a process called intramembranous ossification. A typical long bone in figure 4.4 consists of the following structures:

- Epiphysis — ends of long bones secondary centers of ossification
- Diaphysis — shaft of the bone primary center of ossification

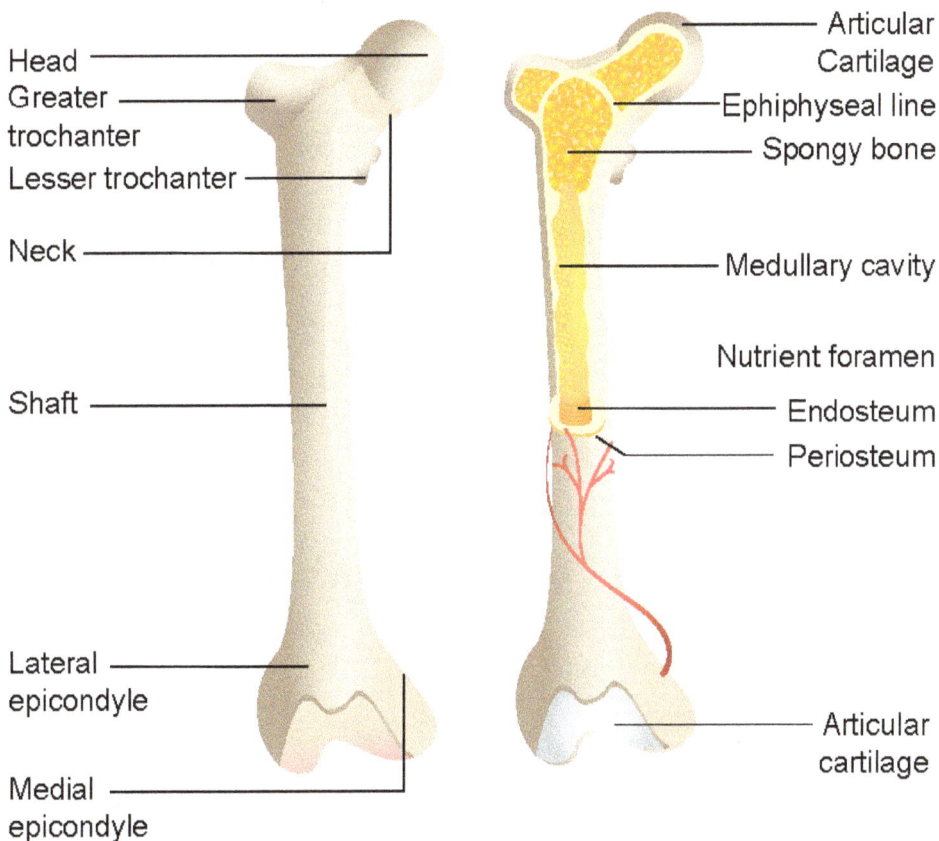

FIGURE 4.4
Typical long bone with cutaway view.

- Metaphysis — the area where the diaphysis and epiphysis fuse
- Medullary cavity — the yellow marrow cavity within the center of a long bone
- Nutrient canals — holes in bone for passage of blood vessels.

The human skeleton (fig. 4.5) is composed of 206 individual bones. The system is divided into two parts the appendicular and axial divisions. The appendicular skeleton is composed of the shoulder girdles, pelvic girdle, the upper extremities and the lower extremities. In this division of the human skeleton there are a total of 126 bones. The axial division is composed of the skull (cranium and face), the vertebral column, and the ribs and sternum. The axial division is composed of a total of 80 bones. Figure 4.5 gives a general breakdown of these two divisions of the skeletal system and accounts for the 206 bones.

In order to be specific as to the actual naming of structures within the human body there are terms that are used to describe sections, directions and in the case of the skeletal system there

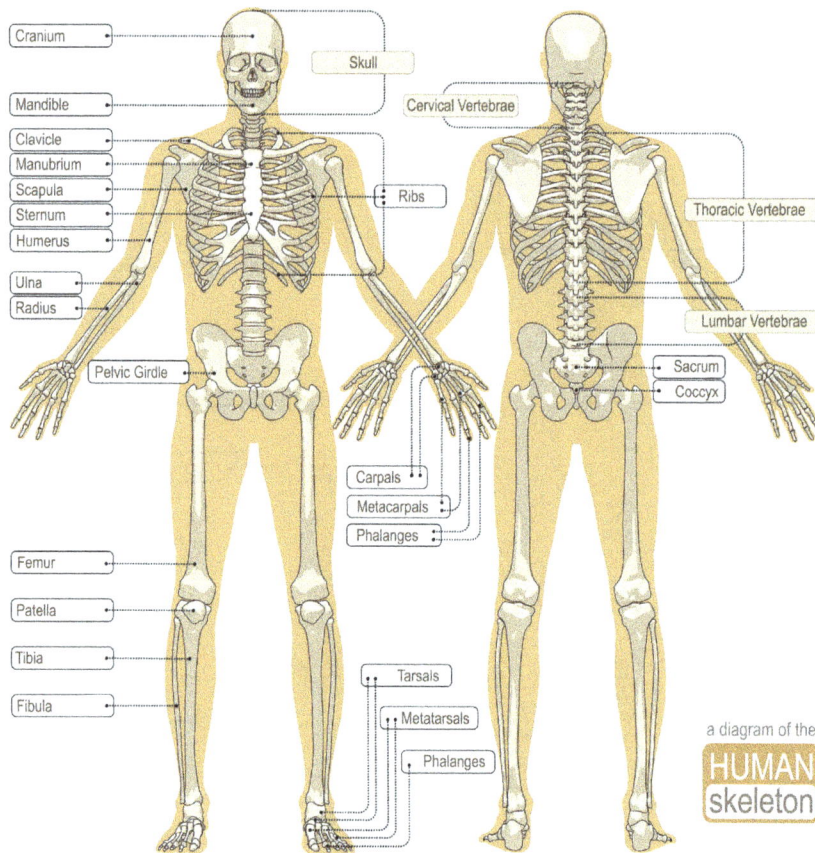

FIGURE 4.5
Axial and appendicular skeleton.

are osteological terms that must be understood. The following lists, tables and diagrams (figures 4.6 and 4.7) are important to the forensic anthropologist and his/her ability to properly identify and characterize the skeletal remains.

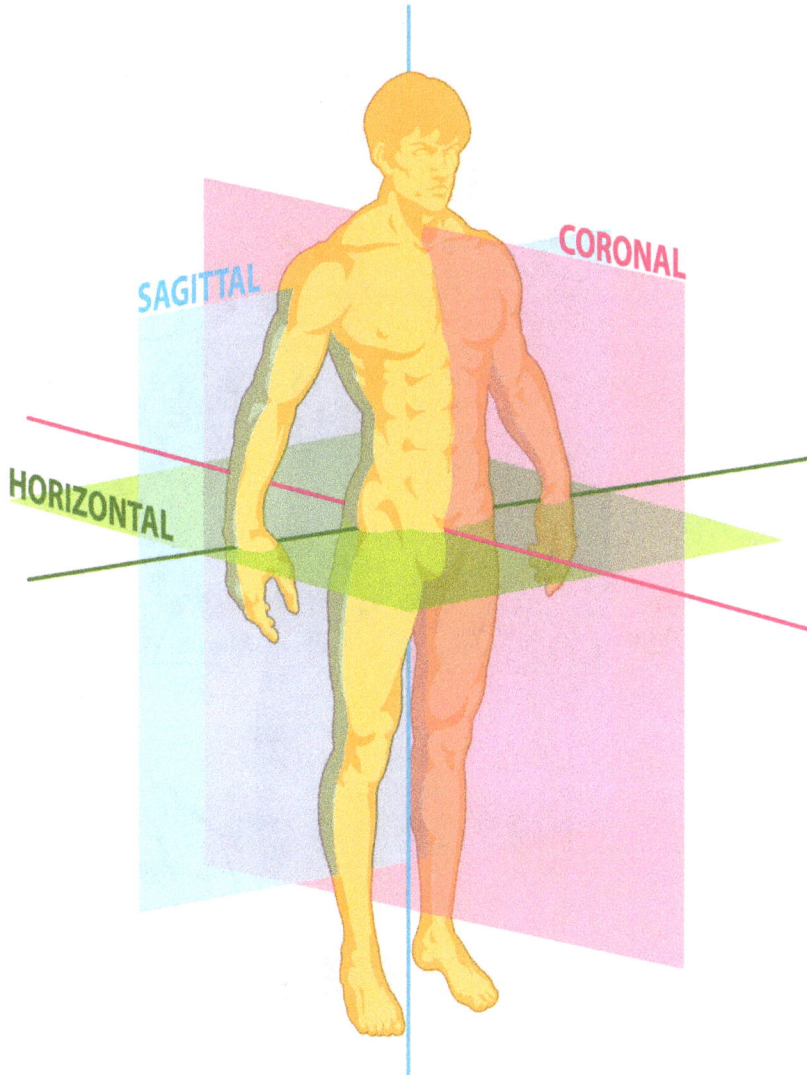

FIGURE 4.6
Body Planes

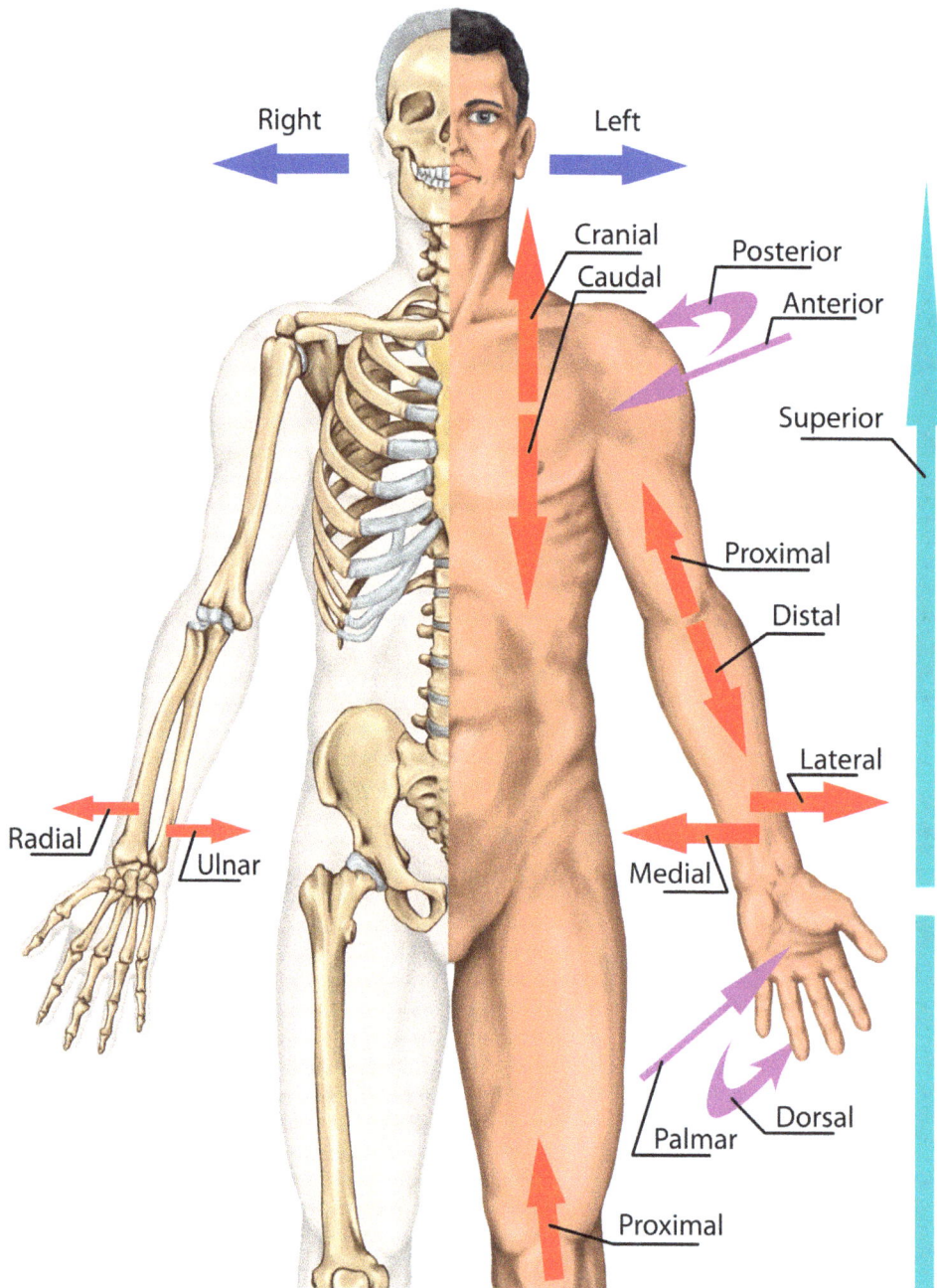

FIGURE 4.7
More Anatomical directions.

Anatomical Directional Terminology:

- **Anterior** in front or in the front part.
- **Anteroinferior** in front and below.
- **Anterolateral** in front and to the side, especially the outside.
- **Anteromedial** in front and toward the inner side or midline.
- **Anteroposterior** relating to both front and rear.
- **Anterosuperior** in front and above.
- **Caudal** below in relation to another structure, inferior.
- **Cephalic** above in relation to another structure; higher, superior.
- **Contralateral** pertaining or relating to the opposite side.
- **Deep** beneath or below the surface; used to describe relative depth or location of muscles or tissue.
- **Distal** situated away from the center or midline of the body, or away from the point of origin.
- **Dorsal** relating to the back; posterior.
- **Inferior** (infra) below in relation to another structure; caudal.
- **Ipsilateral** on the same side.
- **Lateral** on or to the side; outside, farther from the median or midsagittal plane.
- **Medial** relating to the middle or center; nearer to the medial or midsagittal plane.
- **Posterior** behind, in back, or in the rear.
- **Posteroinferior** behind and below; in back and below.
- **Posterolateral** behind and to one side, specifically to the outside.
- **Posteromedial** behind and to the inner side.
- **Posterosuperior** behind and at the upper part.
- **Prone** the body lying face downward; stomach lying.
- **Proximal** nearest the trunk or the point of origin.
- **Superficial** near the surface; used to describe relative depth or location of muscles or tissue.
- **Superior** (supra) above in relation to another structure; higher, cephalic.
- **Supine** lying on the back; face upward position of the body.
- **Ventral** relating to the belly or abdomen.
- **Volar** relating to palm of the hand or sole of the foot.

Examples of directional terms:

- The head is superior to the heart
- The fibula is lateral to the tibia
- The leg is inferior to the thigh

- The brain is deep to the skull
- The skin is superficial to muscle
- The arm is proximal to the forearm
- The hand is distal to the shoulder
- The left ear and left eye are ipsilateral to each other.
- The ears are contralateral to each other
- The heart is medial to the arms

Anatomical Section Terms:

- **Abdominal** — relating to the abdomen. The abdomen is the part of the trunk between the chest and pelvis. It can be divided into three regions: the front, the belly; in back the loins; and on the sides, the flanks.
- **Antecubital** — region of the arm in front of the elbow
- **Brachial** — over the brachial artery in the upper arm
- **Buccal** — of or relating to the cheeks or the mouth
- **Calf** — of or relating to the calf
- **Femoral** — relating to the femur or thigh
- **Inguinal** — the groin or area in lower lateral regions of the abdomen
- **Lumbar** — area over the lumbar spine
- **Popliteal** — region on the back of the knee
- **Scapular** — of or relating to the area near the shoulder blade (scapula)
- **Umbilical** — relating to the central area of the abdomen near the bellybutton

The following lists are all useful in differentiating male from female skeletons, determining race (figure 4.10) and calculating stature. Skeletal remains have been studied extensively after WWII, the Korean War and the Vietnam War. The Central Identification Laboratory (CIL) in Hawaii is charged with the identification of skeletal remains from past wars and tragedies. Much of the research that has been done in this area was conducted back in the sixties and seventies as a result of the government trying to identify skeletal remains of war dead.

Examining Male and Female Skulls for Sexual Differences

1. First note the differences in overall size, shape, and rugosity.
2. Then compare foreheads; how large is the supraorbital ridge?
   How sharp is the orbital rim?
   Are there bumps on the frontal bone?
3. Next compare the mandibles:
   Is the chin squared or oval?
   Can the flare of the mandible be seen?

4. Now, turn the skull and compare the facial profiles:

Shape and contour of the forehead?

How large is the brow ridge?

5. Look at the skull where the ear was:

How sharp is the angle of the mandible? Is it flared?

How large is the mastoid process? Where does the zygomatic arch end?

6. Finally, compare the cranial bases:

Are the nuchal ridges rough or smooth?

How large is the occipital protuberance?

Observing the human skeleton at various ages shows significant differences in skeletal biology. When the age is sub-adult there are distinctive differences in long bone length as is evidenced by looking at the fetal skeleton. There are also distinctive differences in dental eruption and presence of a particular dental profile. The fetal skeleton in figures 4.8 and 4.9, the relatively large amount of cartilage that is present especially the fontanels within the skull plates.

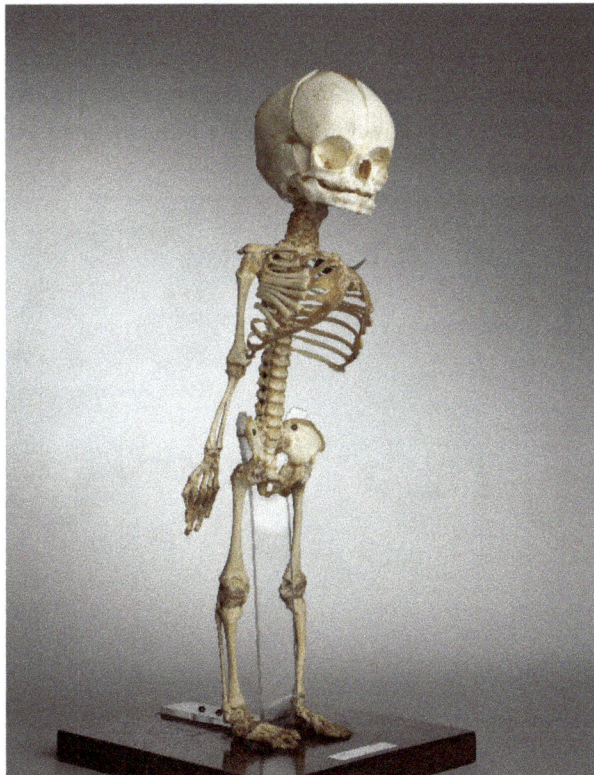

FIGURE 4. 8
Fetal skeleton.

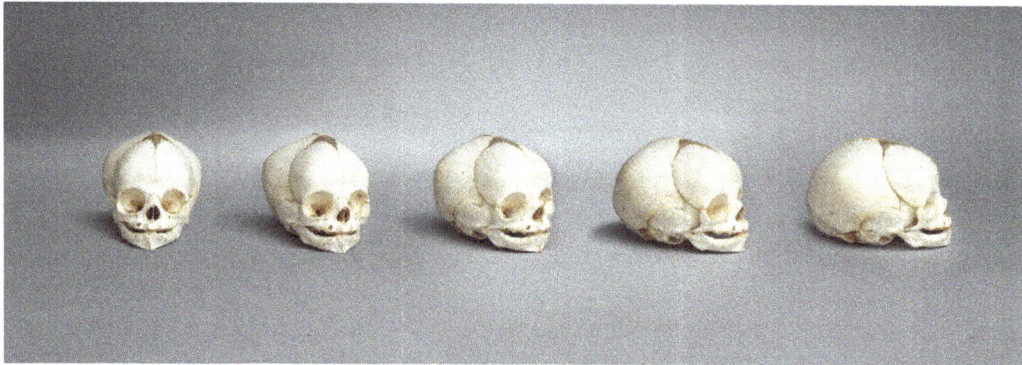

FIGURE 4. 9
Fetal skulls.

Sub-adult skeletons show differences in epiphyseal closure which is a means of determining the age of a human skeleton.

The medical examiner works with the forensic anthropologist and the major aspects of their work is to become involved in skeletal recovery, which might be in the field and the forensic autopsy which is usually conducted in the laboratory. The primary purpose of studying the skeletal profile is to determine age, race, sex and stature which may lead to a positive identification. Also it is noted whether or not there was any bone trauma which might be indicative of foul play.

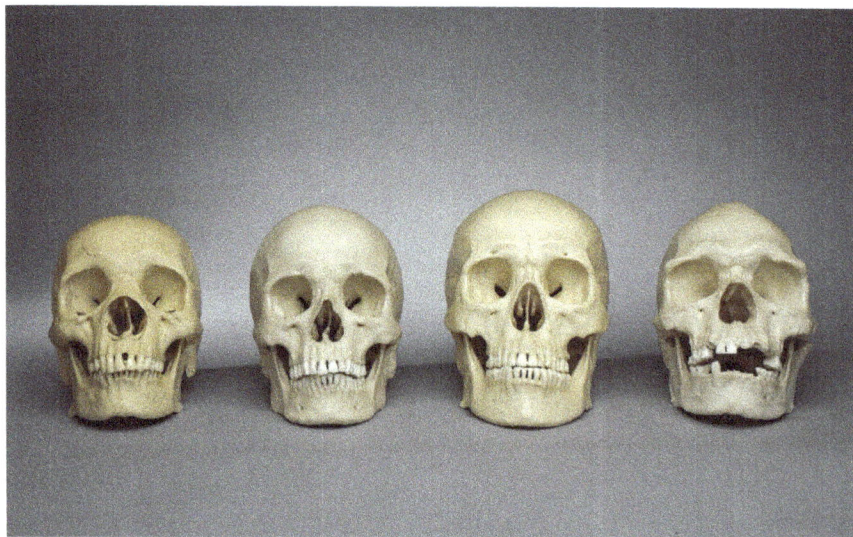

| African Skull | Caucasian Skull | Asian Skull | Aboriginal Skull |

FIGURE 4. 10
Racial features of human skulls.

The most difficult part of the analysis is to try and attempt to determine the time since death. Time since death and the estimation of this period may be divided into stages the first of which is the early stage and includes factors such as algor mortis, livor mortis, rigor mortis, skin slippage discoloration, marbling and odor. Mid state evens may involve the analysis of soil and entomological activity. Late stage changes involve studying botany, bone weathering, carnivore activity. The accumulation of these changes which occur after death are referred to as forensic taphonomy.

Rates of decomposition are dependent on depositional influences such as the depth of burial, surface vs. deep, water content of the surrounding soil and whether or not the body was in water. Factors that accelerate decay or decomposition would be things such as trauma, infection and cycles of freezing and thawing. Actions that would decay decomposition would be factors such as freezing, tannic environment and anaerobic environmental conditions. If trauma is noted then it must be determined whether or not the trauma was ante mortem, perimortem or postmortem.

In many areas of the world today there are mass graves being identified as a result of political and military violence. The Red Cross and many other humanitarian organizations are involved in opening these mass graves and identifying the skeletal remains which is usually all that is found because of the years of time involved between the times of the deaths and the discovery of the remains. Major excavation missions have been carried out in many parts of the world including Argentina, Guatemala, Bolivia, Brazil, Croatia, El Salvador, Ethiopia, Honduras and Iraq. In the search and exposure of these mass graves it is very important to make a determination of how many individuals were in the gravesite. This is done by carefully and accurately collecting the

FIGURE 4. 11
Mass grave site.

skeletal remains and making an accounting of every bone. There is an index called the MNI (minimal number of individuals) which helps determine the exact number of people who were buried. Why bother to determine the minimum number of individuals (MNI)? MNI may be one of the only results possible. Under such conditions, MNI can be the one critical piece of physical evidence that supports or refutes verbal testimony. Pictured in figure 4.11 is a mass gravesite where workers are carefully exposing and removing all skeletal elements possible. Any skeletal element lost or ignored lessens the probability that the individual will be identified. Identifications have resulted from finding nothing more that the stump of an amputated fingertip, a benign tumor on a toe, or the bones of an unusually arthritic neck.

FOR FURTHER READING

Harmon, K. (2009, July 10). Forensic anthropologists aim to identify bodies in cemetery scam. *Scientific America*. Retrieved from http://www.scientificamerican.com/article.cfm?id=foresnic-cemetery-scam

1. The major challenge to the forensic anthropologist is to determine the _____ of the individual whose skeletal remains have been recovered.

2. The upper and lower extremities are part of the _____ skeleton.

3. The cell within compact bone that helps maintain the deposition of calcium and phosphate salts is the _____.

4. One main differentiating factor of the male pelvic girdle is that the pubic arch is _____ 90°.

5. A very wide nasal aperture is characteristic of a(n) _____ skull.

6. In relationship to the elbow, the wrist would be found in the _____ direction.

7. A plane that divides the body into equal right and left sides would be the _____ plane.

8. An abbreviated or shortened mastoid process would be that of a _____ skull.

9. The name of the bone found in the upper part of the arm is the _____.

10. The basic structural unit found in compact bone is called the _____ system.

11. Flaring of the ramus of the mandible is characteristically seen in the _____ skull.

12. The total number of major bones in the human skeleton is _____.

13. The central elongated shaft of a typical long bone in the human skeleton is called the _____.

14. In directional terms, the heart is _____ to the stomach.

15. Skeletal remains represent the end stage of the process of _____.

FORENSIC ODONTOLOGY

LEARNING OBJECTIVES

After completion of this chapter, students will be able to:

- » Define the science of forensic odontology
- » Construct a model of a human tooth
- » Identify the structural components of a human tooth
- » Differentiate an incisor, canine, and molar tooth type
- » Number the teeth in the human maxillae and mandible
- » Evaluate the characteristics of a bite mark
- » Compare the bite marks from various animals

Odontology is the branch of medicine that deals with the study of the development and diseases of the teeth. A forensic odontologist is primarily concerned with using teeth as a means of identification. Very often in mass disasters, earthquakes, fires, bombings and such the remains of individuals may be limited to a few fragments of bone and possibly some teeth and no soft tissue from the body. The forensic odontologist is trained to utilize teeth as a means of determining the identity of an individual when all that can be found are teeth. Through the use of dental records, x-rays and specific bite marks a positive identification can usually be made.

The anatomy of a tooth would be an essential area to begin our study of forensic odontology. Human teeth are covered with a component of enamel which is the hardest material found in the human body. This is one of the main reasons that teeth are often the only elements remaining after a disastrous event. They are relatively impervious to physical forces such as pressure and heat and are able to withstand most environmental stressors. Table 5.1 introduces some terminology which is important in assessing structural characteristics of teeth in relationship to direction and orientation within the oral cavity.

There are three general classifications to human teeth which are incisors, canines, and molars. The sketch below (figure 5.1) shows the anatomy of human canine type of tooth. This is the typical structure for all of the adult teeth. Teeth begin to develop during the early embryo and small tissue regions called buds eventually develop into the deciduous and permanent dentition.

This sketch was submitted by a former student by the name of Corey Ryan. Teeth are organs being composed of a variety of different tissues. The dentin within the center of the tooth is

Table 5. 1   Dental terminology.

Directional Terms Teeth		
Term	Definition	Opposite
apical	toward the root tip	incisal or occlusal
buccal	surface toward the cheek	lingual
cervical	around the base of the crown (neck)	none
distal	away from the midline of the tooth	medial
facial	toward the lips or cheek	lingual
incisal	toward the cutting edge of anterior teeth	apical
interproximal	between adjacent teeth	none
labial	surface toward the lips	lingual
lingual	surface toward the tongue	labial or buccal
mesial	toward the midline of the mouth	distal

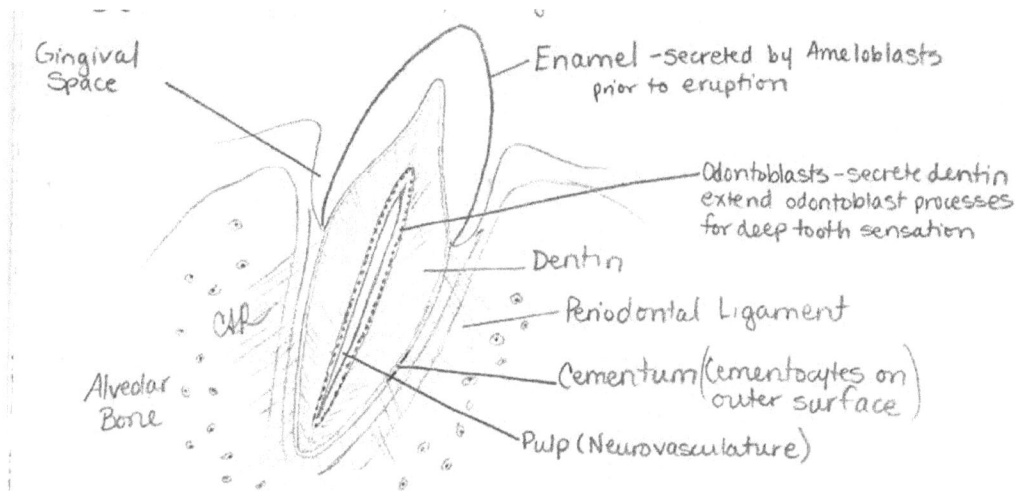

Handwritten labels on the figure:

Gingival Space

Enamel – secreted by Ameloblasts prior to eruption

Odontoblasts – secrete dentin extend odontoblast processes for deep tooth sensation

Dentin

Periodontal Ligament

Cementum (Cementocytes on outer surface)

Pulp (Neurovasculature)

Alveolar Bone

PL

FIGURE 5.1

Anatomy of a canine tooth.

called dentin and is secreted by cells called odontoblasts which line the outside of the pulp cavity. Before human teeth erupt through the gum they are covered with an outer layer of cells called ameloblasts which secrete the enamel. Once the tooth has erupted the ameloblasts break off and the tooth is left with its full component of enamel projected above the gum line. The surface of each tooth is unique to the individual who possesses that tooth and the dental patterns of individuals vary much like individual fingerprints.

Dentition develops over the first 18–20 years of life and is characterized by phases. In the diagrams below these phases are delineated and the eruption of the deciduous (baby teeth) and permanent teeth are presented in a chronology of months and years. In the sequence of diagrams the deciduous teeth are shaded and adult or permanent teeth are white.

As you can see from the previous sequence there are rather marked changes in the dental pattern between the ages of birth and adult. The forensic odontologist must be able to study a mandible or maxillae and determine the age by observing the unique dental pattern. The two diagrams (figure 5.2 and 5.3) demonstrate the deciduous and permanent dentition with the upper jaw (maxillae) and the lower jaw (mandible) open. The dental pattern is described by a numbering system (figure 5.2) which identifies the last permanent molar on the upper right side as tooth #1. Continuing around the teeth on the maxillae from right to left we count from #1 to #16. Then going down to the lower left third molar and giving it the number 17 and continuing around the front of the mandible until we end on the lower right side with tooth #32. This is standard practice in dentistry and this is the way a dentist would identify your teeth.

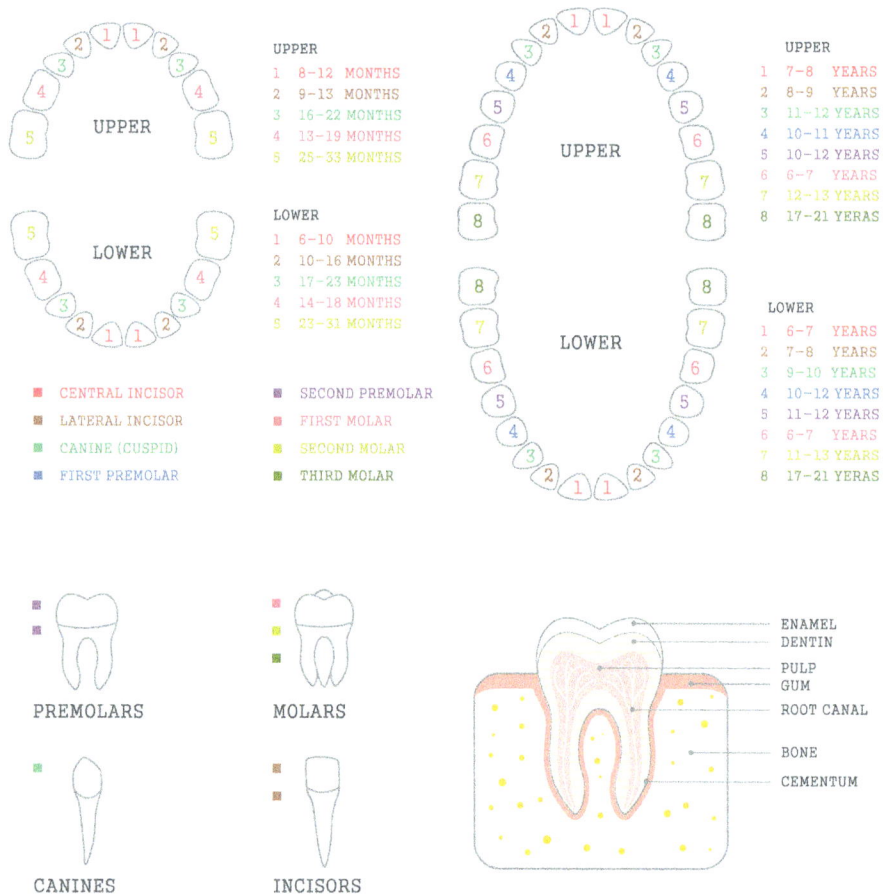

UPPER

1	8-12 MONTHS
2	9-13 MONTHS
3	16-22 MONTHS
4	13-19 MONTHS
5	25-33 MONTHS

LOWER

1	6-10 MONTHS
2	10-16 MONTHS
3	17-23 MONTHS
4	14-18 MONTHS
5	23-31 MONTHS

UPPER

1	7-8 YEARS
2	8-9 YEARS
3	11-12 YEARS
4	10-11 YEARS
5	10-12 YEARS
6	6-7 YEARS
7	12-13 YEARS
8	17-21 YERAS

LOWER

1	6-7 YEARS
2	7-8 YEARS
3	9-10 YEARS
4	10-12 YEARS
5	11-12 YEARS
6	6-7 YEARS
7	11-13 YEARS
8	17-21 YERAS

■ CENTRAL INCISOR   ■ SECOND PREMOLAR
■ LATERAL INCISOR   ■ FIRST MOLAR
■ CANINE (CUSPID)   ■ SECOND MOLAR
■ FIRST PREMOLAR   ■ THIRD MOLAR

PREMOLARS   MOLARS

CANINES   INCISORS

ENAMEL
DENTIN
PULP
GUM
ROOT CANAL
BONE
CEMENTUM

FIGURE 5.2

Deciduous and permanent dentition.

Note that there are only 20 teeth in the complete deciduous or primary individual.

In addition to dental patterns and individual teeth be used as a means of identification they are also useful in identifying an individual who has committed a crime and bitten the victim. Perhaps one of the most famous cases of a serial rapist/killer, Theodore Bundy utilized bite mark impressions on the victim to corroborate the evidence in the case. In many recent cases bite marks on the skin and foodstuffs have proven to be important items of evidence for convicting defendants in homicide and rape cases. If a number of points of similarity can be shown to exist between a bite mark and an individual's dental pattern a forensic odontologist may be able to conclude that and individual is guilty or not guilty of the offense. Bite marks

CHILDREN TEETH ANATOMY

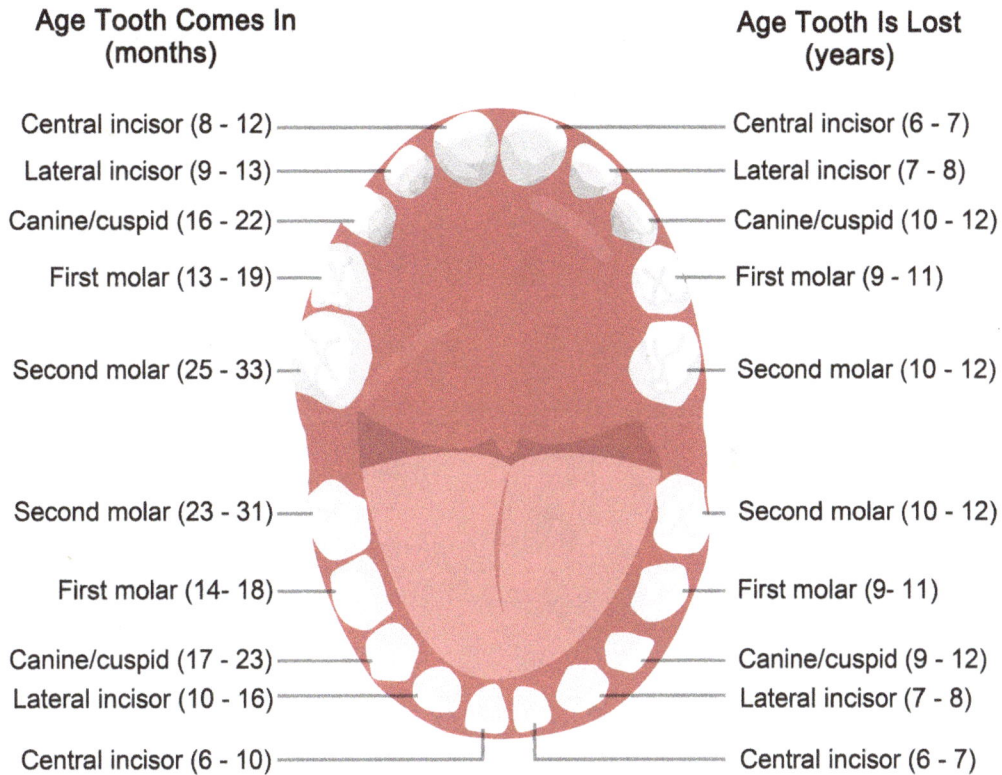

Age Tooth Comes In
(months)

Age Tooth Is Lost
(years)

Central incisor (8 - 12) —————— Central incisor (6 - 7)

Lateral incisor (9 - 13) —————— Lateral incisor (7 - 8)

Canine/cuspid (16 - 22) —————— Canine/cuspid (10 - 12)

First molar (13 - 19) —————— First molar (9 - 11)

Second molar (25 - 33) —————— Second molar (10 - 12)

Second molar (23 - 31) —————— Second molar (10 - 12)

First molar (14- 18) —————— First molar (9- 11)

Canine/cuspid (17 - 23) —————— Canine/cuspid (9 - 12)

Lateral incisor (10 - 16) —————— Lateral incisor (7 - 8)

Central incisor (6 - 10) —————— Central incisor (6 - 7)

FIGURE 5.3
Teeth anatomy in children.

may also be useful in linking an individual to a scene by looking for bite marks on inanimate objects such as chewing gum or a piece of food that has been eaten. Look at the picture of the apple to the right (figure 5.4) and you can easily identify some very distinctive dental patterns. In some instance the forensic odontologist is asked to determine whether or not a bite is from a human source or some other animal such as shark bites or dog bites (figure 5.5) as seen in the photo.

Degenerative changes in teeth are much more difficult to assess than formative changes. As with all organ systems in the body these degenerative changes are related to diet, and individuals' nutritional state as well as the general state of health of the individual. In our more modern day of dentistry many degenerative changes do not occur as early as in the

FIGURE 5.4
Bitten apple.

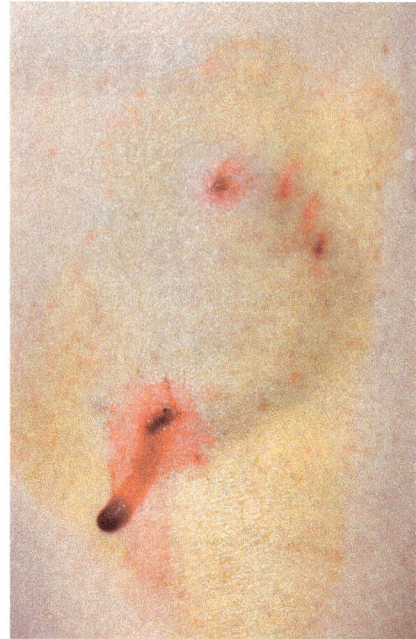

FIGURE 5.5
Bite mark on victim.

past largely because of better professional dental hygiene. Some changes in aging may be observed on x-rays but ground sagittal sections of undecalcified teeth are recommended for viewing age-related changes. This was first suggested by Gustafson in 1950 and it has been tested and modified several times since then by Burns and Maples 1978–79. The following list describes some of the major degenerative changes that occur during the aging process and it is relative to the conditions we described above regarding overall stat of health and physical condition.

- Attrition — loss of tooth crown due to abrasion
- Secondary dentin — deposition of minerals within the pulp chamber
- Periodontosis — apical migration of the periodontal attachment level
- Root transparency — sclerosis of the root dentin beginning with the apex
- Cementum deposition — thickening of the cementum layer
- Root resorption — resorption and flattening of the apex.

The table below (Table 5.2) presents scoring information for age-related data from teeth. There are various regression formulae that relate the various scores, states with and estimated age.

In the various categories of A, S, P, T, C, and R the higher total stage score you register is associated with your age.

Table 5. 2 Age related data from teeth.

Score	Stage 0	Stage 1	Stage 2	Stage 3
(A) crown attrition	no attrition	attrition into enamel only	attrition into dentin	attrition into original pulp chamber
(S) secondary dentin	no secondary dentin	secondary dentin visible	secondary dentin filling 1/3 of the pulp chamber	secondary dentin filling most of the pulp chamber
(P) periodontosis	periodontal attachment at C-E junction	reduced periodontal attachment	periodontal attachment at the upper 1/3 of root	periodontal attachment at the lower 2/3 of the root
(T) root transparency	no transparency	beginning transparency	transparency of the apical 1/3 of the root	transparency of the apical 2/3 or more of the root
(C) cementum	thin, even cementum	increasing cementum	thick layer of cementum	heavy layer of cementum
(R) root resorption	no resorption and open apex	beginning resorption and closed apex	flattening of root apex, affecting only cementum	flattening of root apex, affecting both cementum and dentin

FOR FURTHER READING

Santos, F. (2007, January 28). Evidence from bite marks, it turns out, is not so elementary. *New York Times*. Retrieved from http://www.nytimes.com/2007/01/28/weekinreview/ 28santos.html

1. The branch of science that deals with the study of teeth is _____.

2. The upper dental arcade of an adult contains a total of _____ teeth.

3. The teeth in a toddler that will eventually be lost to permanent teeth are called _____ teeth.

4. Animals that have primarily canine type teeth are considered to be _____ in relation to their diet.

5. The type of tooth that is designed for cutting is called a(n) _____.

6. The enamel surface on human teeth is secreted by _____ before teeth erupt through the gum.

7. The forensic odontologist is primarily concerned with the _____ of an individual through analysis of dental patterns.

8. Bite marks on human skin are sometimes difficult to identify because of the _____ nature of the skin.

9. The term used to describe the portion of the tooth seen above the gum line is the _____.

10. Through the sequential numbering of the teeth in the upper and lower quadrants, the tooth that is number 3 would be classified as a(n) _____.

DERMATOGLYPHICS: SKIN, HAIR, PRINTS, FIBERS, AND BURNS

LEARNING OBJECTIVES

After completion of this chapter, students will be able to:

» Identify the major regions of a section of human skin
» Label a diagram of a cross section of human skin
» Identify the various types of fingerprints and ridge characteristics
» Compare the various structures of human and animal hairs
» Differentiate various fibers and identify them
» Evaluate thermal burns and classify them as first, second, or third degree
» Apply the rule of 9s in calculating the percentage of surface area burned

The word dermatoglyphics comes from the Greek words (derma, meaning skin) and (glyphe, meaning carve) and refers to the so called friction ridge formations which appear on the palms of the hands and the soles of the feet. Typically these areas are free of hair which makes the ridge patterns visible and recordable. The ridge formation of the skin of an individual begins to form during the third and fourth month of fetal development. Contact between the membranes of the amnion and the walls of the uterus influence the development of the specific dermal ridge patterns (dermal pegs) in addition to the influences of genetic makeup. During the lifetime of an individual one has a unique set of dermal ridge characteristics which are valuable for purposes of identification. After death decomposition of the human skin has a later onset in these areas of the hands and feet which provides for a longer time period for making a positive identification. There have actually been cases where the dermal ridge patterns have been the only identifiable part of a deceased individual. There are some records that show the presence of ridge patterns on the hands of 2000 year old mummies from Egyptian anthropology. The human skin, hair and fingernails comprise a very complex organ system. Certain characteristics of this organ system have valuable use in the area of forensic biology. Any transfer print from hands, fingers, toes, ears (on a pillow for example) may provide information as to the identity of the individual who left the transfer print. Secretions from both sebaceous and sudoriferous glands in the dermis of the skin allow for the transfer of these dermatoglyphic markings. Any scaring of the skin leaves an indelible mark which may be recognized for identity. Tattoos, birthmarks and other skin alterations such as surgical incisions, moles, warts or other growths may lead to a means of positive identification of an individual. The human skin plays an integral role in homeostatic regulation for the human body especially in relation to protection to underlying tissues, thermoregulation, and vitamin D synthesis. Forensic biologists are interested in the skin primarily as a means of identification through the observation of transfer of fingerprints and other body surface areas. Additionally the forensic specialist is concerned with the human skin in relation to a variety of burn patterns associated with thermal, chemical and abrasive burns. A following description of the human skin is essential for the forensic biologist to be able to understand the nature of its use in forensic investigations.

The diagram below (figure 6.1) shows the general anatomy of the skin including many of the histological features associated with the skin. The skin is the largest organ in the human body and may comprise about 1.5–2.0 square meters of surface area. A breach to this intact surface area may result in a series of events which could lead to death and it is here that the forensic biologist needs to be able to interpret the findings.

As seen in the diagram (figure 6.1) the skin is divided into three major regions including the outer most layer the epidermis which sits on top of the dermis and beneath the dermis is the subcutaneous or hypodermic region. There are five layers of cells which comprise the epidermis: the bottom most layer is the stratum basale which is made up of the germinal epithelial cells which divide and move upwards to the top layer. Above the basale is the stratum spinosum,

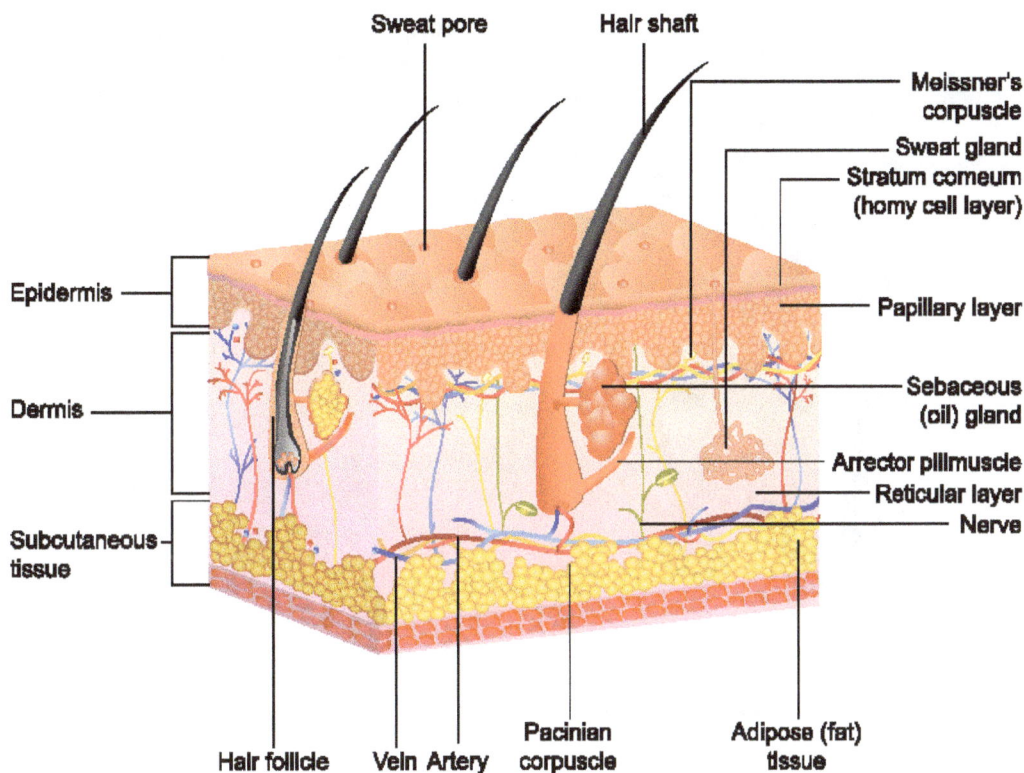

FIGURE 6.1
Skin structure.

the stratum granulosum, the stratum lucidum (in thick skin only) and the outermost layer of stratum corneum. The turnover rate for this tissue is very rapid (a matter of a few days) and there is a continual sloughing off of the outermost layer of the stratum corneum.

Within the dermis are a number of specialized structures in association with the dense irregular connective tissue fibers (composed of collagen) including the hair follicles, sebaceous glands, eccrine sweat glands, blood vessels, nerves, specialized sensory receptors such as Pacinian and Meissners corpuscles and at the upper most surface of the dermis are the dermal papillae which are related to the dermal ridges on the exterior surface of the fingers and toes. The subcutaneous (hypodermis) is largely composed of adipose tissue (fat) and a number of blood vessels and nerves.

At the beginning of the forensic autopsy the pathologist examines the body surface for lacerations, punctures, contusions (bruises) and any other marks which might indicate potential homicide. In cases of strangulation petichiae may be observed in the sclera of the eye or possibly in the oral cavity. In addition to examining the skin as a means of determining cause of death the pathologist will also identify specific markings such as scars, tattoos and birthmarks as a means for identification.

FIGURE 6. 2

Aboriginal indian petroglyoh.

Fingerprints

Fingerprints are of great importance as a means of positive identification. Dermatoglyphics, the science that deals with the observation and classification of hand and footprints, as well as other surface patterns of the skin, dates back centuries. Cave drawings and petroglyph diagrams dating back thousands of years. The figure 6.2 shows a pre-historic drawing. The significance of these pre-historic samples is subject to broad interpretation.

What can be stated with certainty is that as early as 500 BC Babylonian business transactions are recorded in clay tablets that include fingerprints and at approximately the same time ancient Chinese documents and coins are found as seen below (figure 6.3).

Perhaps the most bizarre use of fingerprints in recorded history dates to sixteenth century China where the sale of children is concluded by placing their hand and foot prints on the bill of sale. The first "official" mention of fingerprints is in 1684; Dr. Nehemiah Grew lectures to the Royal College of Physicians of London about the interesting markings found on human fingertips. The next two centuries finds scientists busy exploring the globe, cataloguing animal and plant species, and learning about the basic form and function of the human body. During this period, the study of fingerprints and line formations inches forward. The three most common

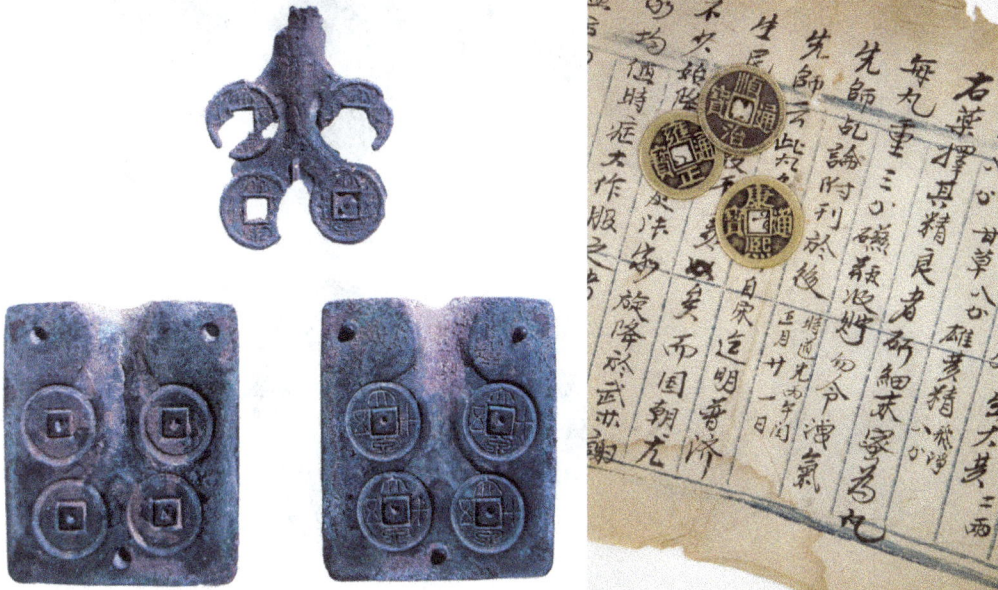

FIGURE 6. 3

Ancient Chinese coins and documents.

print patterns are the loop, arch and the whorl. Each of these ridge characteristics is fairly easy to discern as seen in the pictures in figure 6.4. Additional ridge arrangements are shown along with the three most common types.

A more detailed study of these patterns will be conducted in the lab so that each student will be aware of their own print patterns and how they become specific identity markers for the individual. In addition to the general ridge characteristics there are detailed specific line formations within the print which are useful for detailed differentiation. On the following page (figure 6.5) are some examples of these characteristics such as forks, bifurcations, short ridges, islands, deltas and ridge endings. They are listed in order of how common each is seen in the print patterns of humans.

Hair

Hairs are considered as epidermal derivatives as they arise from the base of the hair follicle which forms from an invagination of the epidermis. The hair grows out of the hair follicle from the root or bulb which is embedded in the base of the follicle and continues to grow into a shaft and terminates at the tip. The shaft of the hair is the part most observed by forensic biologists and it is composed of three layers as seen in figure 6.6.

FIGURE 6.4

Fingerprint patterns; loops and whorls.

FINGERPRINT RIDGE CHACTERISTICS

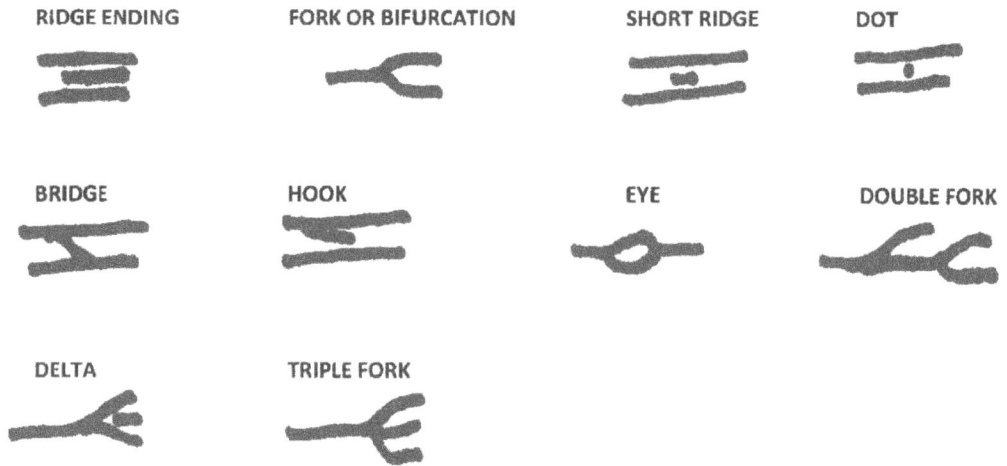

RIDGE ENDING

FORK OR BIFURCATION

SHORT RIDGE

DOT

BRIDGE

HOOK

EYE

DOUBLE FORK

DELTA

TRIPLE FORK

FIGURE 6.5

Fingerprint Ridge Characteristics.

HAIR ANATOMY

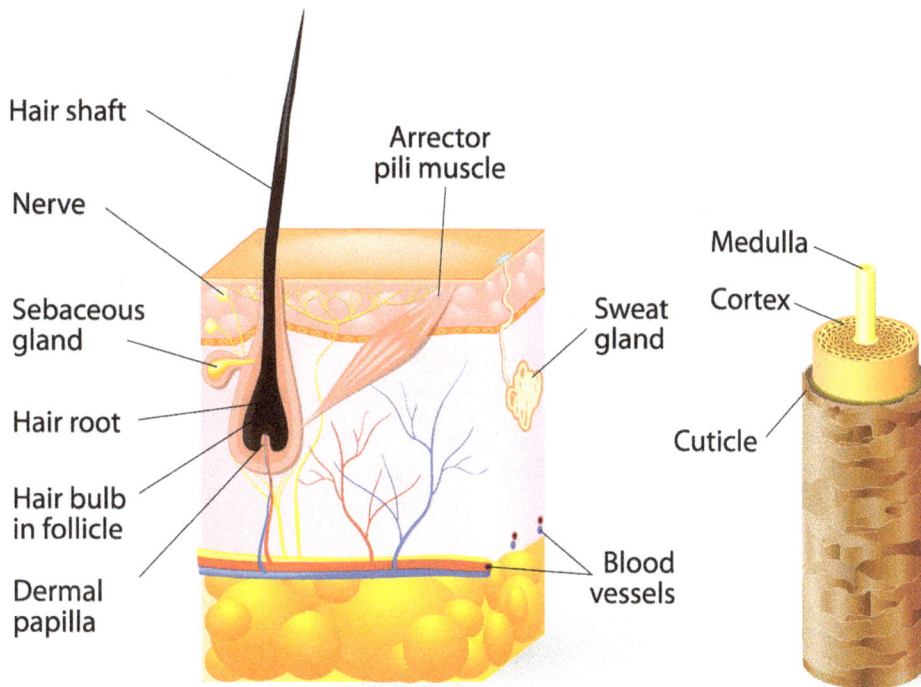

Hair shaft
Arrector pili muscle
Nerve
Sebaceous gland
Hair root
Hair bulb in follicle
Dermal papilla
Sweat gland
Blood vessels
Medulla
Cortex
Cuticle

FIGURE 6.6
Hair shaft.

A more detailed view of the hair follicle is seen in figure 6.6 and although the forensic biologists mainly interested in the shaft of the hair it is noteworthy to comprehend how the hair follicle and associated structures are arranged. In the lower right is a microscopic section taken through some human skin. Note that the hair follicles are found in the lower part of the dermis very close to the hypodermis. The arrector pili muscle which attaches to the connective tissue root sheath is responsible for the onset of shivering thermogenesis which is an early physiological mechanism for increasing body temperature.

Hair is one of the most common types of physical evidence and is considered class evidence. In the best case scenario a human head hair has odds of 4500 to 12 of originating from the same person or 0.0022%. Pubic hair has even better odds of 800 to 1.3 or 0.00012% or originating from the same person. Negroid and Mongoloid hairs tend to exhibit less variation and therefore usually lower odds. Because of these odds, although only a class characteristic, hair can be highly distinctive evidence. The roots of hairs go through growth phases where the initial growth phase is referred to as the anagenic phase which later on when the hair is mature it is referred to as the telogenic phase. Hair roots may be differentiated between various mammals, such as humans (figure 6.7), cats, and dogs. A variety of scale patterns that appear on the cuticle of the hair in the form of coronal, spinous and imbricate. These patterns are detected in the lab through the use of a latex casting of the whole hair. Medullary patterns and types are also distinctive features of hairs. These structures are readily evident by observing the hair under the compound light microscope.

Fibers

Fibers are often the most common type of evidence found at a crime scene. The importance of fiber identification is usually magnified in cases of homicide, assault, or sexual offenses. Fiber analysis is accomplished through three processes including: microscopic examination which includes identification of color, texture, shape pattern, twist, cross sectional appearance and surface characteristics; chemical analysis to test the changes in fiber type under different chemical conditions such as exposure to acid/base, sodium hypochlorite and organic solvents such as acetone; flame tests to determine the type of burn and residual ash as a result of burning the fibers. These three test procedures will often make it possible for the forensic scientist to identify the origin and manufacturer of the specific fiber and hence be able to match a particular fiber with its source.

SHAPE OF THE HAIR

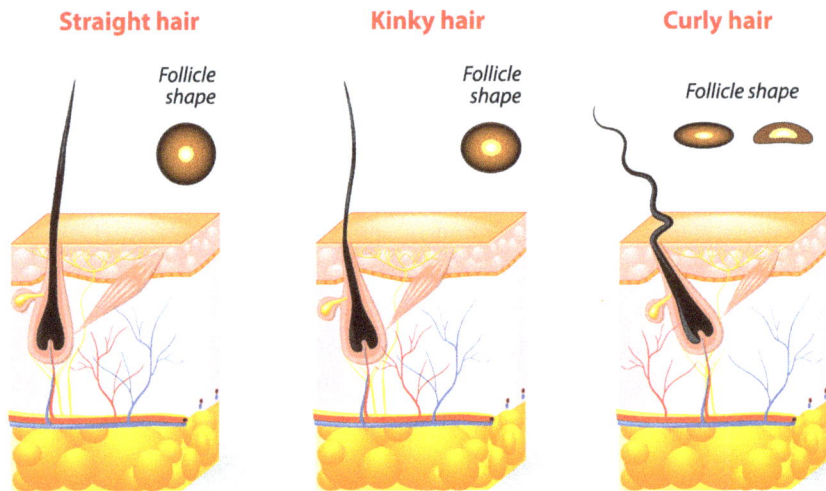

Straight hair　　**Kinky hair**　　**Curly hair**

Follicle shape　　*Follicle shape*　　*Follicle shape*

FIGURE 6.7
Shape of hair.

There are two broad groupings of fibers which include natural and man-made. Natural fibers include those materials that are derived in part or in whole from animal or plant sources such as wool, cotton and silk. Man-made fibers are manufactured synthetic and include nylon, polyester, acetate, and rayon. Fibers also take on very specific color patterns when they are reacted with certain dyes. The observation of fibers after they have been subjected to various dyes as well as an analysis of the particular dyes in the fabric are useful ways of identifying fiber types. Fibers may also be evaluated by determining melting point, density, and tensile strength, however, these tests will destroy the fibers.

Burns

In forensic investigations that involve arson, automobile accidents and chemical accident burns are often evaluated for cause and severity for purposes of treatment and cause of death in a fatal situation. Burns may be caused by intense heat (fire), caustic chemicals such as acids and bases, as well as those caused by friction. The classification of burns is based largely on the assessed damage caused to the layers of the skin — epidermis, dermis and subcutaneous. They are recorded as first, second and third degree depending on the depth of deterioration of the skin. First-degree burns damage the epidermis only and usually produce local edema. Second-degree burns involve damage of both the epidermis and the dermis. Third-degree burns involve the epidermis, dermis and the subcutaneous layers of the skin and are the most damaging which requires immediate care and treatment.

The major consequence of burns is to manage infection and fluid loss. As soon as the epidermis is compromised the possibility of invasion with pathogens will result in infection which can be life threatening by itself. Another consequence of destroying the epidermis is that the body will begin to lose large volumes of fluid. Fluid loss in large percentages of surface area burned may shift the fluid balance and cause vascular problems. To accurately determin the percentage of surface area burned medical personel use the rule of 9's as a quick reference to determine percent surface area burned, to what degree and how much fluid must be replaced to maintain the homeostasis of the body's fluid compartments. Figure 6.8 shows the normal layers of the skin and the involvement of 1st, 2nd and 3rd degree burns.

The treatment of burns is a very complex intervention primarily because of the incidence of infection and fluid loss. Once the outer layer of skin, epidermis, has been removed the underlying tissue becomes susceptible to any and all of the normal pathogens found in the external environment. An intact epidermis prevents the invasion of pathogens and sets up a normal flora and fauna on the stratum corneum of the epidermis which minimizes invasion by foreign pathogens. The intact skin also prevents fluid loss and once the epidermis is violated, tissue fluid is lost quite rapidly. So in the treatment of burns one primary concern is to keep the surface clean and provide an aseptic environment on the burn surface. This is done using sterile

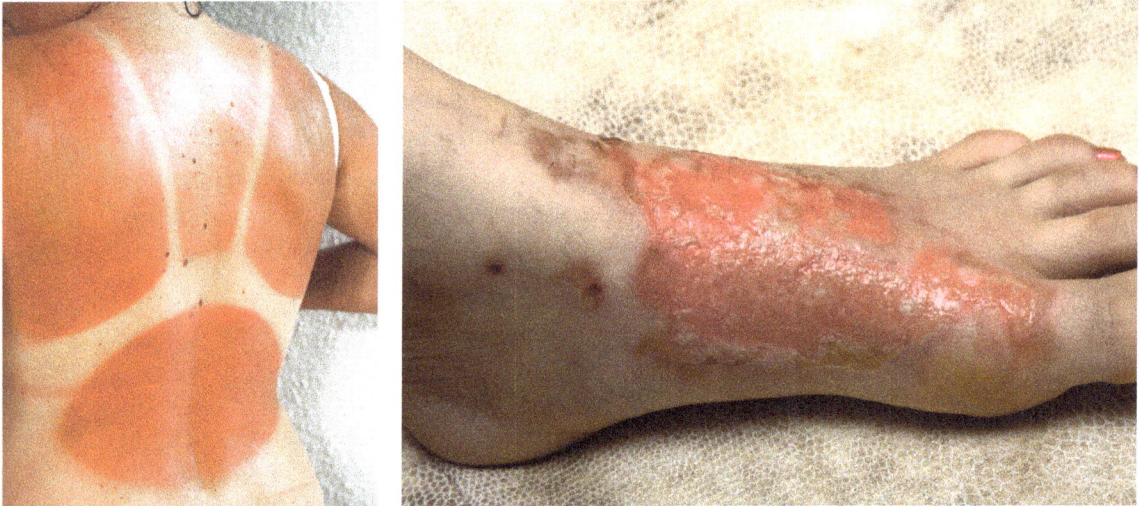

Layers of the skin and the involvement of
1st, 2nd, and 3rd degree burns.

Skin Burns

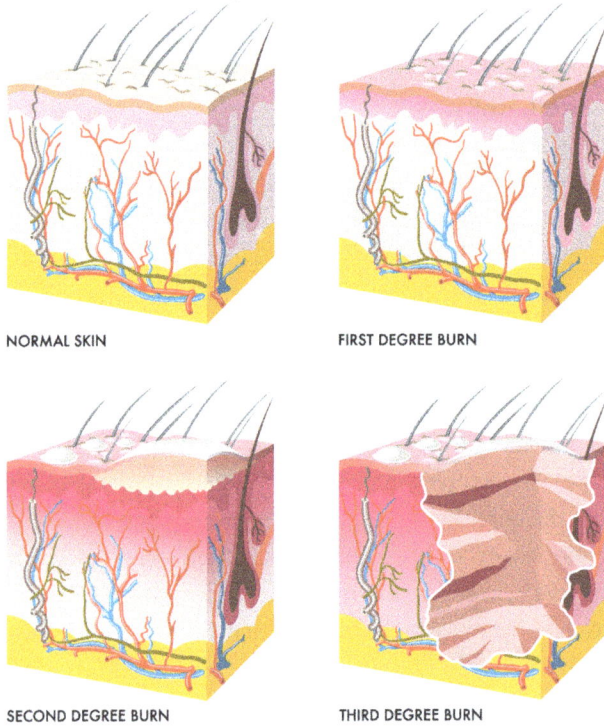

NORMAL SKIN

FIRST DEGREE BURN

SECOND DEGREE BURN

THIRD DEGREE BURN

FIGURE 6.9

Sample of Skin burns.

techniques, using solutions, dressings and eventually perhaps artificial skin or skin grafts. The fluid loss parameter is also important and fluid replacement is related to percent surface area burned. A reference model for calculating percent surface area burned is referred to as the rule of 9's. Each region of the body, head, limbs, front and rear trunk, legs and perineum comprises a percent of the body surface which is a factor of nine. The relationship of the surfaces on the body and the rule of 9's is seen in the model in figure 6.10. The planning of fluid replacement for burn victims is based on the Parkland formula (4 ml lactated ringers/kg/% TBSA burned). In order to effectuate the fluid volume being replaced it is important to calculate the percent TBSA burned. Half of the value is given over the first 8 hours after the time of the burn (not from time of admission to ED) and the other half over the next 16 hours. Therefore a 70 kg man burned over 50% of his TBSA would need to receive 4ml. x 70kg. x 50% = 14,000 ml in 24 hours. This value will be measured against the urine output and the blood volume.

RULE OF 9's

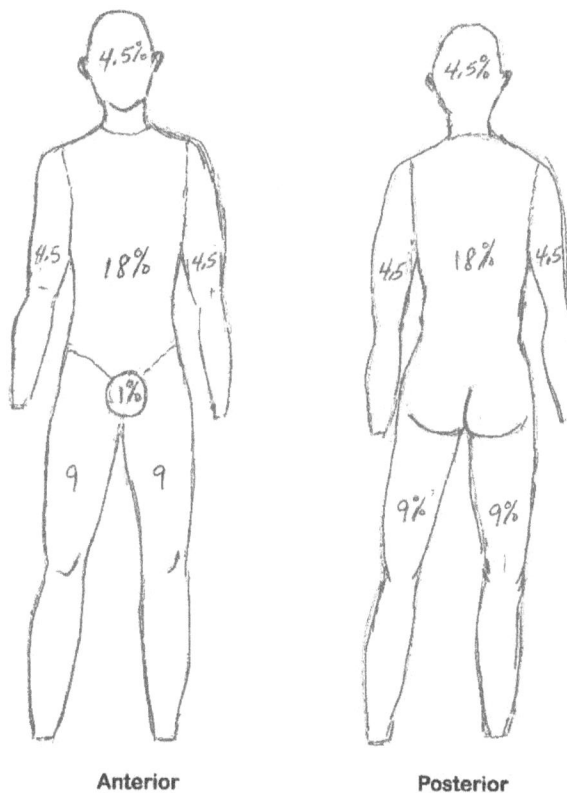

Anterior   Posterior

FIGURE 6.10

Relationship of the surfaces on the body and the rule of 9's.

FOR FURTHER READING

Hamacher, B. (2012). Couple charged after child's severe burns go untreated. NBC 6 South Florida. Retrieved from http://www.nbcmiami.com/news/Couple-Charged-After-Childs-Severe-Burns-Go-Untreated-138957359.html

1. The outermost layer of human skin is called the _____.

2. The three basic types of fingerprints are called _____, _____, and _____.

3. Thermal burns on the skin that would cause blisters and redness are classified as _____ degree burns.

4. An example of a natural fiber is _____.

5. An example of a synthetic fiber is _____.

6. The fingerprint patterns that develop on the distal regions of the fingers are called _____.ridges.

7. The natural oils of the skin are secreted by _____ glands found in the dermis.

8. The outer region of a human hair is called the _____.

9. The root of a hair would yield live cells from which _____ could be found to aid in identification.

10. According to the rule of 9s, someone burned on the total surface of both legs would have _____% of the surface area of their body involved.

11. An internal ridge characteristic in which the ridges branch apart is called a(n) _____.

12. Fluid replacement is important in the treatment of burns, and approximately _____ mL of lactated ringers solution are replaced per kilogram of body weight/% surface area burned.

13. One of the major functions of the skin is to protect the underlying tissues and prevent the entrance of _____ that might cause infection.

14. The average-size individual weighing 70 kg has approximately _____ to _____ square meters of skin.

FORENSIC DNA ANALYSIS

LEARNING OBJECTIVES

After completion of this chapter, students will be able to:

» Construct a three-dimensional model of DNA
» Explain the concept of complementary base pairing
» Contrast a nucleotide to the full structure of the double helix
» Explain the work of Kary Mullis and Alec Jeffreys
» Describe the concept of DNA fingerprinting using RFLPs and STRs
» Identify a match in the DNA fingerprint
» Explain the use and function of restriction endonucleases

Deoxyribonucleic Acid, DNA, the genetic material found in the cells of all living organisms has become a significant aspect of forensic biology. Since the discovery of DNA by Watson and Crick in the mid-fifties the further understanding of its structure and function through an incredible technological age has led to obtaining DNA samples as a means of positive identification in many forensic cases. Each human being has a rather unique DNA profile and the chances of any two individuals having the same sequences of DNA base pairs are approximately 1 in 6 billion. Considering the world population to approach 6 billion, that would mean that perhaps in the entire world there would be one other individual that would have the same base sequences as you. This is so remote that DNA fingerprinting is the most positive means of identification of an individual that is presently known.

The two major technological achievements that led to the use of DNA in forensic science occurred in the early 1980's. In 1980, Kary Mullis and colleagues at Cetus Corporation at Berkeley, California invented a technique for multiplying DNA sequences in vitro by, the polymerase chain reaction (PCR). PCR has been called the most revolutionary new technique in molecular biology in the 1980's. Cetus patented the process, and in the summer of 1991 sold the patent to Hoffman-Laroche, Inc. for $300 million. In 1984 Alec Jeffreys introduced a technique for DNA fingerprinting to identify individuals. The discovery of restriction endonucleases in certain thermophilic bacteria allowed scientists to cut long strands of DNA into shorter fragments. It was discovered that these sets of fragments from different individuals were markedly different in their size and base pair arrangement. Such variations in DNA are called restriction fragment length polymorphisms (RFLP's) and they are extremely useful in genetic studies and are the foundation for DNA fingerprinting.

Each nucleotide is composed of a 5-carbon deoxyribose sugar, a nitrogenous base (adenine, thymine, cytosine or guanine), and a phosphate radical. The nucleotide units are shown in diagram (figure 7.1).

This arrangement is called complimentary base pairing which states that adenine always bonds with thymine and that cytosine always bonds with guanine. Chargaff's rules state that the A/T ration in DNA is always 1/1 and that the C/G ratio is also equal to 1/1. This bonding forms the basic ladder structure of the molecule where the base pairs represent the rungs of the ladder and the sugars and phosphates form the sides of the ladder. The alternate arrangement of the 5-carbon deoxyribose sugars and the phosphate groups which form the sides of the ladder are bonded together by phosphodiester bonds. The final three dimensional structure of DNA was determined to be that of a double helix which is shown in figure 7.2.

Another unique feature of the DNA molecule is that the hydrogen bonds which link the base pairs A-T and C-G are relatively weak chemical bonds. This means that the double helical structure of the molecule is fairly easy to break apart resulting in the formation of two separate helices. This feature of the DNA molecule is used to run the polymerase chain reaction and through continual opening and closing of the molecular structure scientists are able to amplify

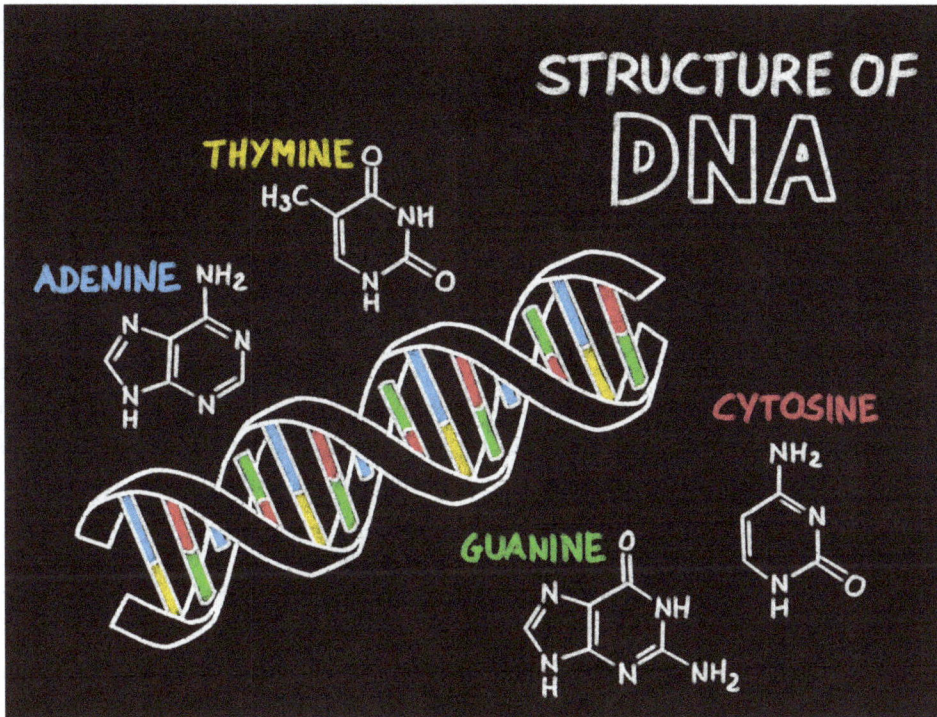

FIGURE 7.1

Base pairs and structure.

DNA replication

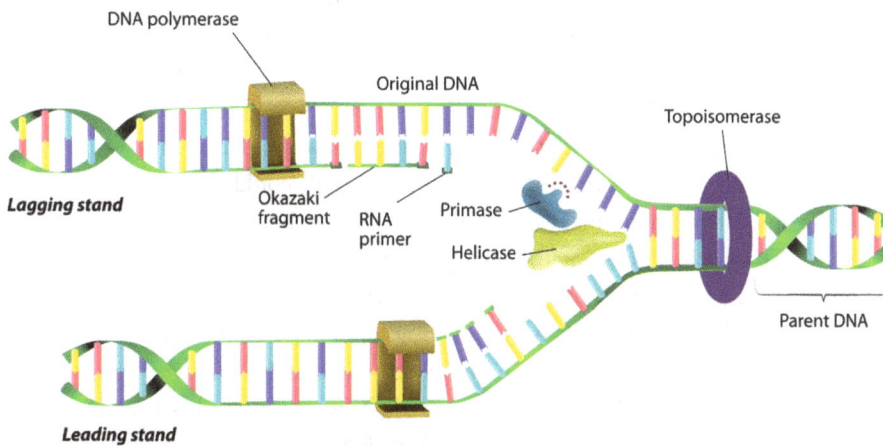

FIGURE 7.2

Three dimensional DNA structure.

a single double stranded molecule into millions in a relatively short period of time. The molecule is also said to be antiparallel in that the phosphate—sugar linkages run in opposite directions on either side of the molecule. Going from the 5' to 3' direction on one side while going from 3' to 5' direction on the opposite side.

DNA Fingerprinting

The concept of DNA fingerprinting is attributable to Alec Jeffreys when he showed that through the use of restriction enzymes the DNA molecule could be cut into many smaller fragments which were unique to each individual. Restriction endonucleases (enzymes) will cut double stranded DNA molecules at very specific locations between certain base pairs. The cut sections of DNA which result from this activity are referred to as RFLP's. There are literally hundreds of different restriction enzymes and by using a variety of these so called "molecular scissors" DNA molecules can be reduced to many different size RFLP's.

Within the DNA molecule itself there are many areas where the nucleotide sequence repeats itself. The regions are called STR's or short tandem repeat sections of the DNA. The process of preparing a restriction digest can be done in the laboratory in a test tube. Through the use of a procedure called gel electrophoresis the individual fragments (RFLP's and STR's) can be loaded into an agarose gel and separated in a buffered electrical field within a gel electrophoresis chamber. The results of this separation show a pattern of DNA fragments that run through the gel and separate based on the charge and molecular weight of each fragment. The greater the molecular weight the more slowly the fragments travel in the electrical field and they lag behind the smaller molecular fragments. The gel on figure 7.3 is that of a DNA fingerprint.

As you can visualize from figure 7.3 the DNA fingerprint is produced from fragments prepared from the DNA of seven suspect's blood and the DNA prepared from a bloodstain typical of that taken from a crime scene. The individual fragments have been separated in the gel based on their molecular weight. You can also see that the fragments separated from the bloodstain appear to match the pattern of suspect 3. At every level the pattern matches between the bloodstain and suspect #3. None of the other suspects have similar banding patterns to those of suspect #3 and even without a trained eye most individuals could read this gel.

One of the newest methods of DNA analysis is using short tandem repeats, STR's, for identification. STR's are loci (locations) in a sample of DNA that contain numbers of repeating patterns. STR sequences are named based on the various lengths of the repeating patterns. Although STR's usually have between 3 -7 repeating units most STR's used in forensic science have tetranucleotide repeats, that is, four bases which repeat over and over. Some examples of forensic STR's include the FGA locus, which has a TTTC repeat and is found in the human alpha fibrinogen locus: the THO1 locus, which has an AATG repeating pattern and is found in the human tyrosine hydroxylase gene: and the TPOX locus, which also has an AATG repeating unit and is located in the human thyroid peroxidase locus.

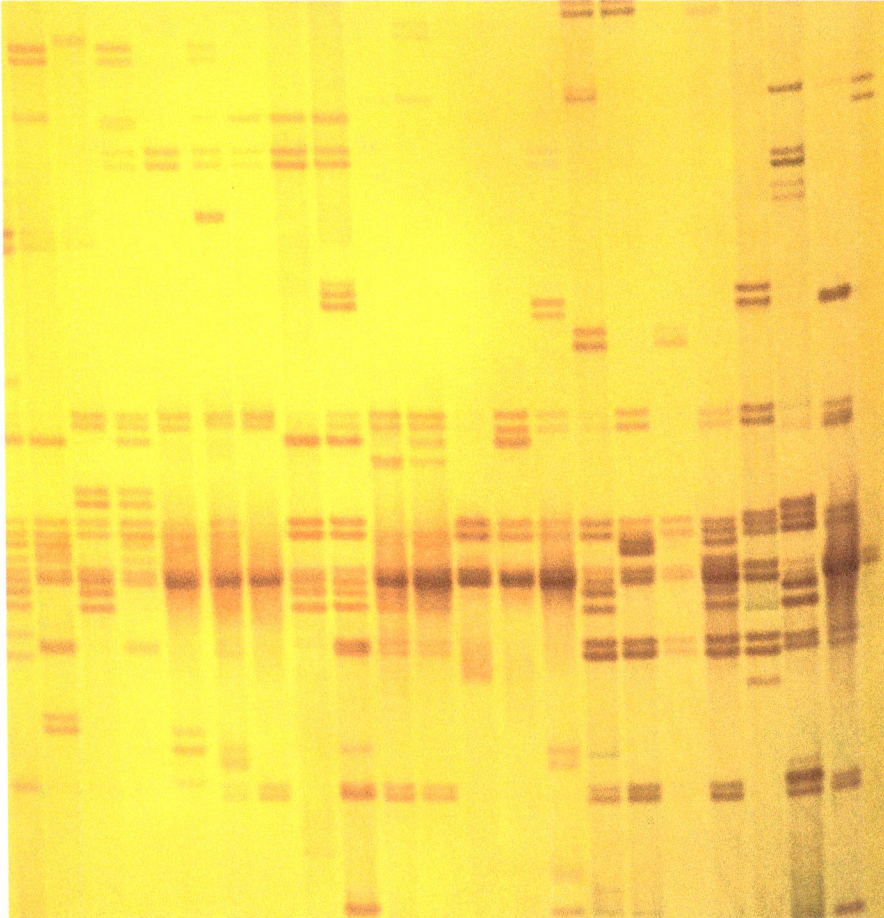

FIGURE 7. 3
DNA fingerprint

The total length of an STR is usually very short (fewer than 500 bases). Due to the extremely short length of the STR it is much less susceptible to degradation than are the fragments used for RFLP analysis. Thus STR's are often recovered from bodies or stains that have started to break down or decompose. The small size of the STR also means that it can be amplified relatively quickly in the PCR amplification reactions which are advantageous when available samples are small.

The number of repeats in STR's varies widely among individuals, which makes them very useful for human identification purposes. For the THO1 STR, for example, the most common repeats are 6-10.

Each individual STR is longer by 4 bases. The 6 allele is 24 bases long, the 7 allele is 28 bases long, the 8 allele is 32 vases long and so on. This fact becomes important when we discuss the separation and identification of each STR.

STR's are scattered throughout the genome and appear, on average, every 10,000 bases. Searches of the recently completed human genome reference sequence have attempted to comprehensively catalog the number of STR's in the genome. The STR markers that are used in forensic analysis are usually chosen from separate chromosomes to avoid any linkage problems.

The alleles mentioned in the section above are inherited from the individual's parents following the fundamental rules of Mendelian genetics. For example looking at figure 7.4 shows an inheritance pattern considering a single chromosome. This particular chromosome has alleles that are STR's and differ in the number of repeats. The father has 2 repeats on one of the chromosomes and 5 repeats at the same locus on the other, so it is said to be heterozygous and has a 2, 5 genotype for this locus. The mother is also heterozygous and has a 3, 8 genotype for the same locus.

During meiosis (figure 7.5), the chromosomes separate, such that each gamete receives only one of the two chromosomes from the parent. In the example above the father's seminal fluid

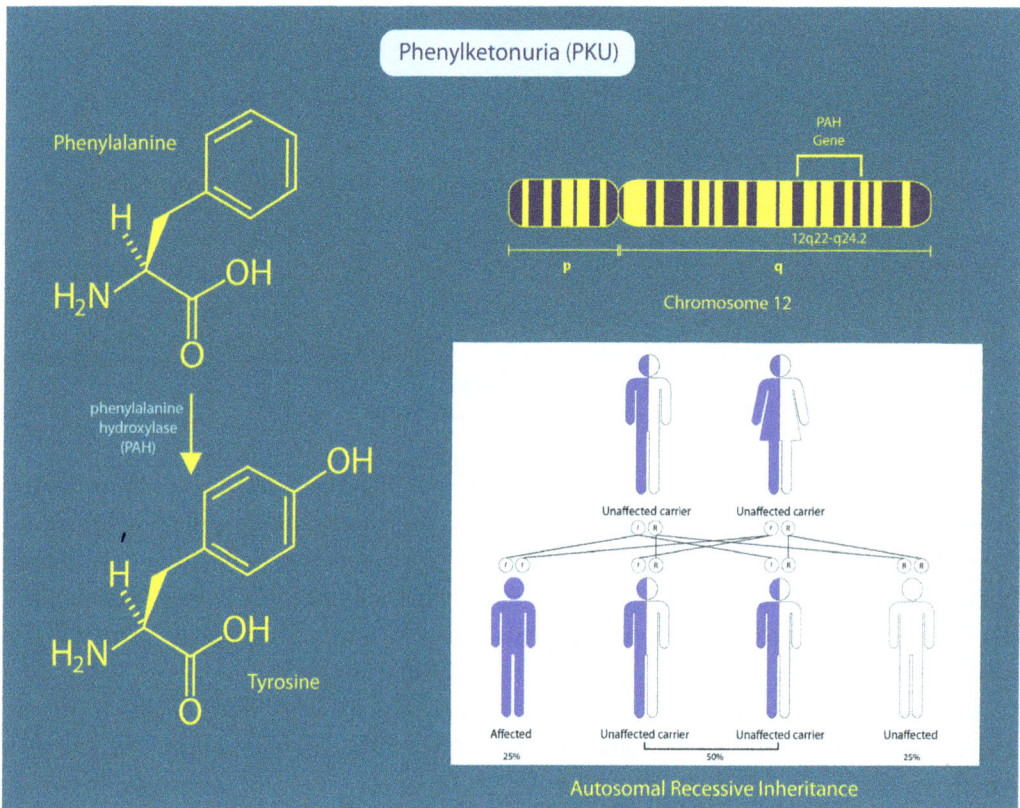

FIGURE 7. 4
Inheritance pattern considering a single chromosome.

Cell division (meiosis)

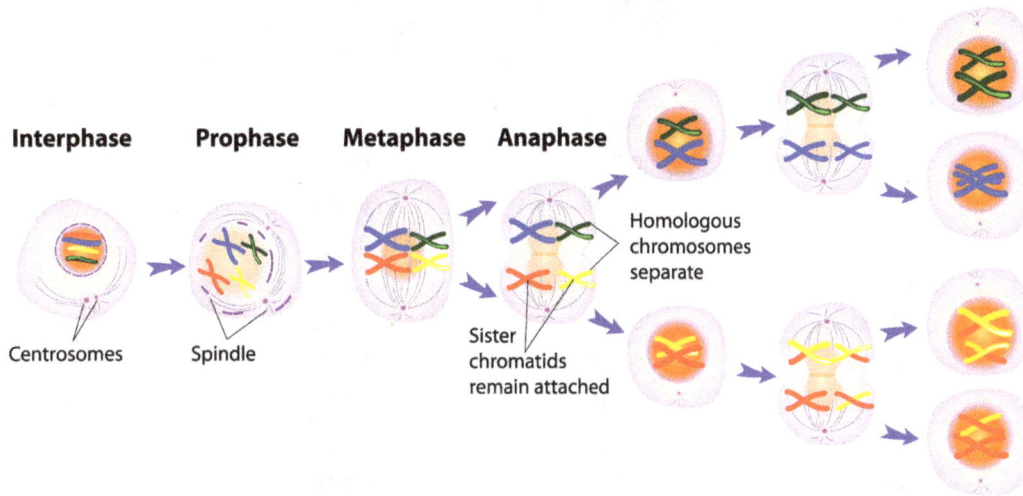

FIGURE 7. 5

Meiosis

will contain a mixture of sperm cells carrying either the 2 repeat or the 5 repeat. The mother's egg cells will carry either a 3 repeat or an 8 repeal allelic format.

At conception, the haploid sperm fuses with the haploid egg, forming a diploid fertilized egg called a zygote. This egg cell will have one of four possible combinations of alleles, and as seen in figure 6 the combinations will be 2:8, 2:3, 5:8, or 5:3. Since these alleles are in non-coding sections of the DNA they will have no effect on the characteristics (phenotype) of the child.

Electrophoresis is a technology used to separate charged particles or ions in solution using an electrical current. DNA molecules, which contain many negatively charged phosphate groups (PO_4-), will migrate through an agarose gel based on size and electrical charge. Since the overall charge on the DNA molecule is negative it will migrate towards the positive pole (anode) within the agarose gel in a phosphate buffered solution. Because the smaller DNA fragments move faster, electrophoresis is used to separate STR's according to their size and charge. The smaller STR's will migrate faster within the gel and the larger STR's will move more slowly hence, separating the STR's into various bands. The DNA samples (STR's, RFLP's etc.) are loaded into wells at the positive electrode (cathode) and they move through the gel, which acts as a sieve, when applying an electrical current of between 50–100 volts. Once separated, the actual STR length (measures in base pairs) is determined by comparison with STR standards. This length can then be used to establish the number of repeats and hence to determine the genotype of the individual at each amplified locus.

In the United States the forensic community has standardized on 13 different STR's which are compiled into the Combined DNA Index System (CODIS). The frequency at which each

Table 7.1   The Thirteen CODIS STRs and Their Probability of Identities

STR	African American	U.S. Caucasian
D3S1358	0.094	0.075
vWA	0.063	0.062
FGA	0.033	0.036
TH01	0.109	0.081
TPOX	0.090	0.195
CSF1PO	0.081	0.112
D5S818	0.112	0.158
D13S317	0.136	0.085
D7S820	0.080	0.065
D8S1179	0.082	0.067
D21S11	0.034	0.039
D18S51	0.029	0.028
D16S539	0.070	0.089

individual STR will occur in the human genome is extremely small and the probability (product of each of the individual frequencies) becomes extremely small between any two individuals. The 13 CODIS STR's are seen in Table 7.1.

FOR FURTHER READING

Fox News Network. (2012). "Who's your daddy?" Paternity testing van offers quick DNA results. Retrieved on from http://www.foxnews.com/health/2012/08/24/who-your-daddy-paternity-testing-van-offers-quick-dna-results/#ixzz26DFFlXoE

1. The three-dimensional structure of DNA is described as a _____.
2. The building blocks of the DNA molecule are units called _____.
3. According to Chargaff's rules, the nitrogenous base adenine also pairs with _____.
4. Kary Mullis revolutionized the use of DNA by developing a technique called _____.
5. The process from question 4 allows for the _____ of very small samples of DNA.
6. The five-carbon sugar in the DNA molecule is called _____.
7. Through the use of restriction, _____ large pieces of DNA may be cut into smaller fragments called RFLPs and STRs.
8. Through separation, using gel electrophoresis, these RFLPs or STRs produce a DNA _____.
9. The chemical bonds that form between the complementary base pairs in the double helix are weak _____ bonds.
10. In 1953 the two scientists who were credited with the discovery of DNA were _____ and _____.
11. DNA is found in the nuclei of cells, and during cell division it is tightly coiled around some histone proteins and exists in the form of structural units called _____.
12. The gametes (sex cells) of an individual contain one half of the genetic material of the normal body cells (somatic cells) and are said to contain the _____ number of chromosomes for that organism.

FORENSIC TOXICOLOGY, ALCOHOL, AND DRUGS

LEARNING OBJECTIVES

After completion of this chapter, students will be able to:

- » Define the science of toxicology
- » Discuss the major routes of entry into the human body
- » Explain the chemistry of carbon monoxide poisoning
- » Classify the general classes of poisonous or toxic substances
- » Give examples of toxins produced by living organisms
- » Explain the significance of determining the LD_{50} value
- » Discuss the major types of alcohols and explain what effects they have on human metabolism
- » Evaluate the types of drugs classified in schedules from the Controlled Substances Act of 1970

Toxicology

Toxicology is the science that deals with the study of poisonous substances. This creates a very broad area of concern because many substances that we might not consider to produce toxic effects, when administered or taken in a reckless manner, may produce such effects. Exactly what is a poisonous substance? By definition any substance natural or synthetic that when introduced into the body causes physiological or psychological effects that directly interfere with the normal metabolic pathways is considered to be a potential poison. The toxicologist needs to be familiar with the chemical composition of such substances and how they interfere with the metabolic pathways where they produce their effects. The primary role of the forensic toxicologist is to investigate and identify toxic substances which might have been associated with a death, to perform postmortem drug testing, occasionally workplace drug testing and identification of contraband materials. Deaths might occur from accidental poisoning, drug abuse, suicidal poisonings or homicidal poisonings.

The toxicological investigation of tissues from the situations mentioned above must adhere to a protocol which allows for rapid and accurate analysis. The studies are performed by collecting samples of all body fluids, including blood urine, saliva and perhaps cerebrospinal fluid (CSF). If the case has resulted in a death tissue samples from all organs might also be analyzed. The forensic toxicologist cannot simply look for the presence of a toxin or a drug in the body but must understand the chemical changes that the toxic substances undergo as a result of their metabolism within the body's organs. Often the original toxin is metabolized rapidly and only its metabolites (end products) may be present after testing. The forensic toxicologist works like a detective as he pieces together all of the substances identified through the testing procedure and uses these identifiable substances to determine the exact chemical nature of the original poison.

An extremely important aspect of determining the nature of a poisoning is to determine the route of entry. There are several routes through which toxic or poisonous substances may produce their effects on the physiology or metabolic activities within the organism. We will classify each route and determine the time frame between administration and possible effects. The oral route of administration, a common route for many over the counter and prescriptive medication, involves entry through the gastrointestinal tract (GI). Some substances, such as nitroglycerine, may be absorbed in the tiny blood vessels under the tongue within the oral cavity. This provides for a fairly rapid route of entry in which the substance may produce its effects within 15–30 seconds. These substances are generally lipid soluble hence the relative rapid absorption. The other important areas within the GI tract where entry may occur would be within the stomach and the small and large intestine. The rate of absorption here might take as long as 15 to 90 minutes. The important point to remember here is that if a drug is administered through the GI tract and it is not absorbed in the oral cavity there is a fairly long time period before absorption begins during which it is often possible to pump the stomach and remove the poison before it is absorbed. The drug IPECAC induces vomiting and provides a means to remove stomach contents, if it is safe to do so. It would not be advisable to evacuate the stomach contents if the ingested material were a caustic substance such the household product Drano which is a strong alkaloid containing sodium hydroxide. Caustic substances will burn the lining of the esophagus during ingestion and would only compound the problem if they were removed through the same lining.

Another route of entry is by inhalation which involves the respiratory system including the nasal passages, trachea, bronchi and the lungs. This route of entry provides a primary means of for gases to enter the body. Substances in the gaseous state will readily enter the body through the respiratory system in as little as a few seconds. The entrance of oxygen (O_2) and the removal of carbon dioxide (CO_2) represent our natural mechanisms for maintaining our normal metabolism. Many anesthetics such as ether and nitrous oxide may enter the body through this route and induce unconsciousness which is necessary during surgical procedures. During these procedures the amount of the gas being inhaled is controlled and produces a state of unconsciousness which can be regulated by an anesthesiologist or a nurse anesthetist. If the inhalation is not controlled such as that occurring when people accidentally or purposely inhale carbon monoxide (CO) the toxic effects of the gas may become lethal. Carbon monoxide is a gas that has 100 times the affinity for hemoglobin than that of preferred oxygen. In the reaction below you can readily see why CO becomes poisonous when inhaled in increased concentrations.

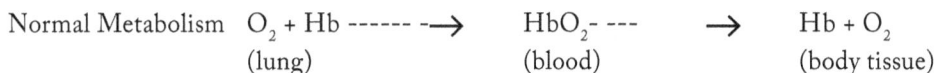

Normal Metabolism O_2 + Hb ------ -$\longrightarrow$ HbO_2- --- $\longrightarrow$ Hb + O_2
                                (lung)   (blood)   (body tissue)

In the situation above oxygen combines with hemoglobin, a protein on the red blood cells (RBC's) in the lung, travels through the blood as a compound called oxyhemoglobin (HbO2) and the oxygen is released within the body's tissues which allows the cells to metabolize the oxygen and produce energy in the form of Adenosine triphosphate (ATP).

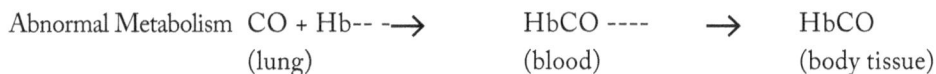

Abnormal Metabolism CO + Hb-- -$\longrightarrow$ HbCO ---- $\longrightarrow$ HbCO
                                (lung)   (blood)   (body tissue)

In this case carbon monoxide (CO) readily combines with Hb (100 times greater than O_2) to form a compound called carbaminohemoglobin (HbCO) and when it arrives in the tissues there is no delivery of oxygen so the normal metabolic pathways that produce ATP are blocked and this results in low energy levels. People who die from carbon monoxide poisoning drift in to a state of low energy followed by unconsciousness and ultimately expire. Going back to our definition of a poison this situation directly interferes with the metabolic pathway that leads to energy production so for that reason it is called carbon monoxide poisoning.

The intravenous route (IV) produces immediate effects and is often used by drug addicts to receive certain opiates and amphetamines. The danger here is that little time is available to reverse the effects of an injection of a toxic substance. In the medical arena it is used to administer anesthetics, fluids and many drugs which are dosed over time. If you have ever had a tooth extraction by an oral surgeon you may have been subjected to the effects of an anesthetic such as sodium pentathol or sodium brevitol usually administered IV. The question asked is that you, the patient, are to count backwards from 100 and as the anesthetic is introduced it is usually 100, 99…maybe 98, you lose concsciousness in approximately 2–3 seconds. The effects of toxic substances introduced through the IV route produce instantaneous effects so if the dose is calculated to be a lethal dose it usually is fatal.

Two other routes, not as common as those described above, are the subcutaneous and dermal routes of entry. The subcutaneous route involves an injection just beneath the dermal layer of the skin. Usually a small, 25 gauge, needle is used to inject a small amount of material to the subcutaneous tissue. The absorption here is relatively slow, perhaps over hours, and is primarily used for detecting certain types of allergic reactions. A dermatologist may inject certain allergens in this area to test a patient for certain allergies such as pollens, animal dander's and food substances. This route is often used to test for the presence of certain antibodies in patients who have been exposed to certain disease causing organisms. A good example is the test for tuberculosis where a small amount of the attenuated organism, *Mycobacterium tuberculosis,* is introduced subcutaneously and after a twenty-four hour period the area is observed for the presence of a small pustule or raised area which if present indicates the presence of specific antibodies within the patient tested. Dermal administration involves a very slow route of entry; perhaps hours to days in which long lasting, usually lipid soluble, and substances may enter the body directly through the skin. In the medical arena chemicals such as nicotine and testosterone may be administered through a dermal patch.

The general classes of poisonous or toxic substances may be grouped under four broad classifications.

1. Gases
2. Metallic poisons
3. Volatile organics
4. Non-volatile organics

There are many substances that exist in the forms of those identified in the outline above many of which may be toxic to the body if administered in a haphazard manner or without proper monitoring of dosage or concentration. The Table 8.0 below lists some examples of the compounds which belong to these groups.

Many toxic substances may be derived from living materials which would include those toxins of plant, animal, microbial and fungal origin. Table 8.1 identifies some of the living materials and what types of toxic substances are produced by each. Some of the most poisonous substances known to man fall within these groups.

There are an array of laboratory methods for detecting the presence of many of these substances. In general the toxicologist uses color tests, micro diffusion tests, various methods

Table 8. 0   Types of Compounds

Class	Examples
Gases	oxygen (O2), hydrogen sulphide (H2S) (carbon monoxide (CO)
Metallic Poisons	mercury (Hg), lead (Pb)
Volatile Organics	methane (CH_4), ether (CH_3OCH_3)
Non-Volatile Organics	many alkaloid substances often derived from plants (amphetamines, cocaine, opiates, cannabinoids)

Table 8. 1   Poisonous Substances

Plants	Animals	Microbial	Fungal
Atropine - nightshade	Tetrodotoxin (puffer fish venom)	Botulinum (*clostridium botulinum*)	Aflatoxins
Coniine — hemlock			Phallotoxins
Ricin — castor bean			Amatoxins (mushrooms)
Digitalis — foxglove	Various arthropods Formic acid (spiders)	Tetanus (*clostridium tetani*)	

of chromatography including thin layer chromatography (TLC), gas chromatography (GC), and high performance liquid chromatography (HPLC). In addition many forms of spectroscopy including UV, visible light, X-ray, infrared, microwave and gas chromatography/mass spectroscopy may be used to isolate and identify very specific chemicals which represent the wide variety of toxic substances. The interpretation of the findings from the various analytical methodologies will always ask the following questions: Is a drug or poison present? What is the substance (chemistry)? How much of the substance is present? How was the drug/poison administered? Is it's concentration in the body sufficient to cause death? The answer to this last question is determined by calculating the lethal dose or the LD_{50} value. The LD_{50} value is defined as the smallest dose of a substance that when administered to a group of test animals (usually mice or rats) will cause death to 50% of the animals. LD_{50} values have been determined for most every compound we can mention. An example of a common LD_{50} value is that for the cannabinoid (marijuana) THC, which by the inhalation route in rats is 42mg/kg of body weight. The lower the LD_{50} value for a substance the greater toxicity it possesses.

Alcohols

Alcohols are a group of organic substances which are often assayed during forensic investigations. Alcohols by definition are those organic chemicals which have hydroxyl groups (OH) attached to a carbon backbone. Three common forms of alcohol are methyl (CH3OH), ethyl (CH3CH2OH) and propyl (CH3CH2CH2OH).

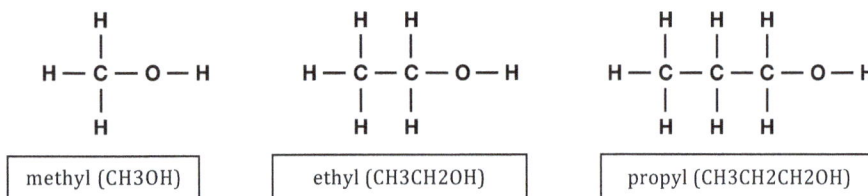

methyl (CH3OH)   ethyl (CH3CH2OH)   propyl (CH3CH2CH2OH)

Methyl alcohol or methanol is known as wood alcohol. It is a byproduct of the distillation of organic material such as wood. It is extremely poisonous and one eight ounce glass could cause blindness and possibly death. Ethyl alcohol or ethanol is an alcohol that is produced by fermenting certain grains and fruits. We consume ethyl alcohol in the forms of various malt beverages, wines and hard liquors. The percent alcohol by volume of each of these beverages is approximately; for one twelve ounce beer 5% alcohol, one six ounce glass of wine 12%–15% alcohol and hard liquor may range between 40%–80% alcohol. The proof label on hard liquor represents a number that is two times the percentage of alcohol by volume. For instance, if the hard liquor is 80 proof the percent alcohol content would be 40%. Blood alcohol levels may be calculated by using instruments such as breathalyzers or simply by a direct measurement from the individual's blood. Most states have set the legal limit for blood alcohol at .08mgs/100 ml. of blood. If your blood alcohol level (BAC) is below .08% you are considered to be in a safe state to operate a motor vehicle. Anything above .08% you are considered to be legally intoxicated and this constitutes a criminal offense if you are operating a motor vehicle or watercraft. Using the following table one can calculate the approximate BAC of an individual at a certain body weight for a certain number of drinks in a specific time frame, examples can be seen in Table 8.2.

Alcohols as a group are fairly lipid soluble and are absorbed in the GI tract in a relatively short period of time. On an empty stomach a small framed individual, five feet two inches and weighing 105 pounds could possibly feel the effects of ethyl alcohol in a matter of minutes.

Metabolism might be defined as a measurement of all of the chemical reactions occurring in the body. Some of the reactions build up materials (anabolic) and some reactions tend to break down substances (catabolic). By understanding alcohol metabolism we can learn how the body eliminates alcohol from the its systems and understand some factors which may affect this process. In addition, understanding alcohol metabolism may also help us understand how this process influences the metabolism of nutrients, drugs (medications) and hormones. Metabolism

Table 8. 2 **Average blood alcohol percentages in people of different weights vs. number of drinks imbibed over a relatively short period.**

Drinks	Body weight in pounds							
	80	100	120	140	160	180	200	240
1	.06	.04	.04	.03	.03	.03	.02	.02
2	.12	.09	.07	.06	.05	.05	.04	.04
3	.18	.14	.11	.10	.08	.07	.06	.06
4	.25	.19	.15	.13	.11	.10	.09	.07
5	.31	.23	.20	.17	.14	.13	.11	.09
6	.37	.27	.23	.20	.17	.14	.14	.10
7		.32	.27	.24	.20	.18	.16	.12
8		.37	.31	.28	.23	.21	.19	.14

** Subtract .01% for each 40 minutes of drinking.
One drink is 1 oz. of 100-proof liquor or 12 oz. of beer.
Example: A 160-lb. man takes 7 drinks in 4 hours (240 min.).
.20 − .06 = .14%

is the body's process of changing ingested substances into other compounds and results in converting some compounds to become more and some to become less toxic than those originally ingested. Metabolism involves many complex chemical reactions and one important one is the process of oxidation. Through the process of oxidation ethyl alcohol is removed from the body's systems by being converted to a less toxic compound. Like many other chemical compounds ethyl alcohol is metabolized in the liver. An enzyme called alcohol dehydrogenase (ADH) is important in reducing ethyl alcohol into a less harmful compound called acetaldehyde. The next metabolic step is to convert the acetaldehyde into acetate by other enzymes and eventually this compound is converted into carbon dioxide (CO_2) and water (H_2O). Most all of the alcohol in the body is metabolized in the liver, however, a small amount remains in the blood and is excreted in the urine and some through the lungs (exhalation).

Many factors affect the absorption and metabolism of alcohol. Food in the GI tract may decrease the absorption of alcohol by a factor of three. Gender is a factor where women have higher BAC's than men after consuming the same amount of alcohol perhaps because women have less body water in their fluid compartments and also have lower concentration of ADH in their stomach contents. Although alcohol has a relatively high caloric value (7 Calories/gram) it does not necessarily relate to weight gain in people who consume large amounts of alcohol. The effects that alcohol produces on the body's systems are seen to occur more rapidly in a lower weight individual. The practical side to understanding how alcohol is metabolized allows us to calculate what our blood alcohol concentration (BAC) would be after drinking considering the factors mentioned above.

Drugs

Drugs possess toxic properties and since they may be natural or synthetic and cause physiological and/ or psychological effects on the body they are may also be considered as poisons. We may categorize drugs as being over-the-counter, prescriptive or illicit. Hallucinogens, amphetamines and barbiturates are drugs. These substances have found their way out of research laboratories and pharmaceutical establishments resulting in common use of many drugs in the general public. In the forensic world most interest in drugs is related to causes of death related to the toxic effects of drugs. In particular illicit drugs, natural or synthetic, become the major interest of the forensic toxicologist.

Many drugs when used in the proper form, proper concentration and properly medically prescribed manner are an important aspect associated with human health. As is the often the case today the improper use of drugs has led to many people developing a physical or psychological dependency, a condition in which the individual experiences and intense physical craving or unreasonable psychological need for the drug. Today we see that the misuse of certain drugs often leads to dependency or addiction. The degree of dependency on certain drugs by an individual is related to many factors, including body weight and overall physical condition of the individual, a particular genetic tolerance for the drug, the environmental conditions which exist at the time of taking the drug, and the very specific characteristics of the drug itself. Addiction is a term which refers to one having a long term dependency that results in an individual becoming irrational in his/her need for a particular drug. Drug abuse may be defined as the non-medical use of any drug in any fashion that is socially unacceptable.

Drug abuse in this country, largely because of the negative effects it has on our citizens, led Congress to pass the Comprehensive Drug Abuse Prevention and Control Act (commonly called the Controlled Substance Act) in 1970. Certain substances that produce physiological or psychological effects on the human body are under regulatory control by the federal government. The basic outline for this governmental control places such substances into five major groups which are referred to as schedules. Table 8.3 outlines the five schedules and provides examples of drugs within each grouping.

Drug Classification

There is a large number and tremendous variety of powders, tablets, capsules, liquids, smoking material, vegetable matter, and related paraphernalia such as cookers, pipes, and syringes that give testimony to the presence and use of illegal drugs in our society. Most controlled substances

Table 8. 3 **Five schedules and sample drug groupings.**

SCHEDULE	SUBSTANCE
I	Substances which have high potential for abuse. No present medical use may be used in research mescaline, lysergic acid diethylamide (LSD), heroin, methaqualone and marijuana.
II	Substances which have high potential for abuse but are currently accepted for medical use. Amphetamines, cocaine, morphine, methadone, phencyclidine (PCP), barbiturates such as secobarfital and pentobarbital and dronabinol the synthetic equivalent of marijuana.
III	Substances currently accepted for medicinal use and show less potential for abuse than those substances in schedule I and II. All barbiturates (except Phenobarbital not covered under schedule II, certain codeine preparations and anabolic steroids).
IV	Substances with a low potential for abuse relative to Schedule III substances, have a medical use [diazepam (Valium), propoxyphene (Darvon), Phenobarbital, meprobamate (Miltown), chlordiaze-poxide (Lithium)].
V	Substances with medical use, low potential for abuse. (Many over the counter preparations, and certain opiate drug mixtures with non-narcotic medicinal ingredients.

contain an active ingredient as well as an additive used to dilute their potency or to enhance their value. When a forensic chemist undertakes the identification of an unknown substance thought to be a drug, his/her investigative techniques must be sufficient to cover all contingencies and to leave no room for error. The validity of his/her results must be able to be supported in a court of law and help to confirm the guilt or innocence of a defendant.

Drugs of abuse may be classified into four separate and distinct groups based on their mode of action on the body's organ systems.

Narcotics/Opiates

A narcotic is any substance that produces a loss of sensations, or narcosis, a stuporous state that resembles sleep. An opiate is any substance, natural semi-synthetic or synthetic in origin, that produces morphine-like effects in the body.

OPIUM

A brown, gummy substance derived from secretions from the seedpods of the poppy plant. Of the alkaloids in opium, which produce the pharmacologic effects, 10% is morphine, 0.5% is codeine. Opium is usually taken into the body by smoking (inhalation route) the substance in long-stemmed pipes.

MORPHINE

An alkaloid found naturally in opium, extracted with solvents or by ion-exchange techniques. The common forms of morphine are morphine sulfate and morphine hydrochloride. These white crystalline powders are water-soluble and have a bitter taste. Morphine may also be found as tablets, capsules, or even in a liquid form. Morphine is usually administered with a hypodermic needle. It causes drowsiness, decreased physical activity, pinpoint pupils, and feelings of tranquility and euphoria. Needle marks are a sign of morphine use.

HEROIN

Heroin is prepared from morphine by a synthesis reaction with acetic anhydride. It is usually found as heroin hydrochloride. This white, crystalline, water=soluble powder is dissolved and administered by injection through the intravenous route (IV). There it is metabolized into morphine. Heroin is usually diluted with quinine, lactose, mannitol, or powdered milk. Heroin is three times as potent as morphine and the euphoria from its use is more intense than that derived from morphine.

CODEINE

Codeine may be extracted from opium or made synthetically from morphine. It is usually found as codeine sulfate or codeine phosphate. Codeine acts as a mild analgesic, having only one-sixth the potency of morphine. Codeine is taken through the oral route of administration and is a common ingredient of cough syrups and pain relief medicines.

HYDROMORPHONE

(Dilaudid) A semisynthetic opiate, usually found in tablet form. This drug has a high potential for abuse.

OXYCODONE

(Percodan) A semisynthetic opiate made by chemically modifying an alkaloid from opium. It produces an effect similar to that of opium.

MEPERIDINE

(Demerol) A totally synthetic opiate, usually encountered in tablet form or as a liquid. Tablets are ingested; the aqueous solution of Demerol is designed for injection. The sedation and euphoria produced by merperidine are similar to the effects of morphine.

METHADONE

(Dolophine) Methadone is as effective and potent as morphine. Methadone is used medically for the relief of pain, and as a treatment for the withdrawal symptoms resulting after chronic use of other opiates.

Depressants

A depressant is any substance that reduces the activity of the central nervous system. This results in the loss of judgment, reduced coordination, inhibition of orderly thought processes, slurred speech and reduced reaction time. High concentrations of depressants in the blood can result in such severe depression of the central nervous system (CNS) which decreases the cardiovascular and respiratory functions that it can lead to unconsciousness or death.

BARBITURATES

Barbiturates are drugs that act on the CNS to produce calming and a feeling of well-being. These drugs, commonly referred to as "downers", are grouped into three categories on the basis of the onset and duration of their actions; short acting barbiturates (pentobarbital, secobarbital), intermediate acting barbiturates (amobarbital), long acting (Phenobarbital).

ALCOHOL

Ethyl alcohol, or ethanol, is a depressant of the CNS. Alcohol is a colorless, water-soluble volatile liquid found in alcoholic beverages. It causes impairment of the ability to walk or drive, slurred speech, and dizziness. Excessively high levels (>0.5mgs) are toxic and can lead to death.

METHAQUALONE

(Quaalude, Sopor) this drug is a nonbarbiturate, sedative-hypnotic drug that depresses the CNS. It induces sleep and acts as a muscle relaxant. Methaqualone is most commonly taken in tablet form.

MEPROBAMATE

(Equanil, Miltown) meprobamate is a nonbarbiturate , sedative-hypnotic drug that depresses the CNS. It is usually taken as a tablet.

DIAZEPAM

(Valium) One of a class of drugs derived from benzodiazepine. It acts to depress the CNS. It is used therapeutically to sedate and to reduce anxiety. The drug is taken most often as a tablet or capsule. High doses of this drug can cause lack of muscular coordination and are substances that decreased reaction time.

CHLORDIAZEPOXIDE

(Librium) One of the benzodiazepine derivatives, it acts on the CNS as a depressant. Abuse of this drug can result in psychological dependency.

Stimulants

Stimulants are substances that heighten the activity of the sympathetic nervous system, which regulates body functions such a as heart rate, respirations, blood pressure, smooth muscle tone, psychomotor activity and appetite level. Stimulants are commonly called "uppers" or "speed". Therapeutic doses of stimulants produce a feeling of well-being and increased alertness. Higher levels of stimulants result in an initial "rush" followed by intense feelings of pleasure and euphoria. Blood pressure is raised, respiratory rate is increased, there is a loss of fatigue, decreased appetite, increased wakefulness, and nervousness or excitement. Excessive use may induce aggressive or psychotic behavior, hallucinations, muscular tremor, and in some cases convulsions and death.

AMPHETAMINES

(Benzedrine, Dexedrine) This drug is often prescribed as an appetite suppressant. It is most often taken orally as a tablet or capsule but also may be found as a liquid or powder. Its use produces an intensely pleasurable, mood-elevating "rush" accompanied by increases in energy, alertness, and enhanced strength. Amphetamines are commonly identified as "uppers" or "speed". Repeated use results in psychological dependency.

COCAINE

An alkaloid obtained from the leaves of the coca plant. Cocaine produces a feeling of euphoria and exhilaration and acts as a powerful stimulant to the CNS. It is ingested most often by inhalation and subsequent absorption of the drug through the mucous membranes of the nasal passages. It has clinical use as a local anesthetic. A form of cocaine known as "crack" is taken into the body by smoking. Use of cocaine and especially crack results in extreme dependency.

CAFFEINE

A mild stimulant most commonly obtained through the drinking of coffee. Abuse results in mild psychological dependency.

NICOTINE

A stimulant found in the leaf of the tobacco plant. Its use causes physical or psychological dependency.

Hallucinogens

Hallucinogens are substances that can cause significant alterations in normal thought processes, perceptions and moods that are identified as psychotic reactions. The effect an individual experiences depends, to a great degree on the personality of the person and the circumstances under which the drug is used. Some users experience a sense of euphoria, colorful visual illusions and apparent distortions of their body parts or objects around them. Others may experience frightening psychotic episodes and paranoid delusions, fear, extreme depression, or manic excitement.

LSD

Lysergic acid diethylamide is a hallucinogenic drug synthesized from lysergic acid, a substance found in a fungus. It produces marked changes in mood, and hallucinations that can last for up to 12 hours. Physical dependency does not develop but "flashbacks" can occur.

PCP

Phencyclidine has anesthetic use in veterinary medicine, but is not used on humans because of it hallucinogenic action. Use of PCP may cause thought impairment, double vision, dizziness, loss of comprehension and of muscular coordination, and disruption of time sense. Use of PCP causes strong psychological dependency. Street names for PCP are "peace", "hog", "pill" and "angel dust". PCP is usually ingested orally as a tablet capsule or powder.

MESCALINE

Mescaline is the primary hallucinogenic ingredient in the peyote cactus. Synthetic mescaline may be made in the laboratory. Mescaline is usually found as a powder and is taken orally mescaline is chemically related to amphetamine.

MARIJUANA

Recently many have questioned the placement of marijuana as a hallucinogen. It has become legal in many states for medicinal use. By federal standards it is still classified into this group. Marijuana is plant material derived from the flower, seeds, stem and leaves of the *Cannabis* plant. The active ingredient of marijuana, with street names of "pot", "grass", "reefer", or "joint" is tetra-hydrocannabinol or THC. Most THC in the *Cannabis* plant is found in the resin, flowers, and leaves of the plant. Little is found in the stem, roots or seeds. Hashish is a resinous secretion from the flowers that has a THC concentration of 5-12%. Liquid hashish is an extract of hashish that has anywhere from 20–65% THC. Marijuana produces a feeling of well-being and a more vivid sense of touch, sight, smell, taste and sound. While use is thought not to cause physical dependency it may cause psychological dependency.

Drug Identification

The first step in the identification of an unknown substance thought to be a drug is to employ a screening test that will reduce the countless possibilities of substances to a manageable number. In the case of pills, tablets, or capsules consult any of a number of books that show pictures of the color, size, shape and marking of current prescription medicines. A good source for this would be the Physicians' Desk Reference, PDR, a yearly publication that identifies all FDA approved drugs. The second step is to subject the unknown substance to analytical tests that positively confirm its identity. In certain situations, the forensic chemist will use a series of non-specific presumptive tests to identify the substance. In other situations, the substance will be subjected to highly specific techniques such as infrared spectrophotometry or mass spectrometry that will identify the substance to the exclusion of all other know chemical substances.

The presumptive tests forensic chemists normally use to identify substances are: color tests, in which a chemical reagent undergoes a particular color change in the presence of a drug; microcrystalline tests that require microscopic examination of the crystals that result from a chemical reaction; chromatography, based on the separation and colorization of the drug (or one of its ingredients) and determination of the Rf values; spectrophotometry which relies on a substance's characteristic absorption of wavelengths of light and mass spectrometry, which relies on the bombardment of a substance with electrons, thereby causing the substance to disintegrate into numerous smaller fragments that are separated by size while passing through a magnetic field. Of these, color tests, microcrystalline tests, and chromatography are the easiest to use.

FOR FURTHER READING

Wenner, M. M. (2012, February 24). Deadly duo: Mixing alcohol and prescription drugs can result in addiction or accidental death. *Scientific America*. Retrieved from http://www.scientificameri-can.com/article.cfm?id=mixing-alcohol-prescription-drugs-result-addiction-accidental-death

1. Any substance that may alter the metabolism of the body is considered to have _____ effects.

2. The route of entry that produces the most immediate effects if a toxic substance would enter the body is the _____.

3. Gasses and vapors usually enter the body through the _____ route.

4. Caffeine, as a drug, would be properly classified as a _____.

5. Carbon monoxide (CO) binds irreversibly with _____ on the red blood cells and has an affinity 100 times greater than that of oxygen.

6. Nicotine is sometimes administered as a treatment through the use of a patch. The route of entry would be described as the _____ route.

7. The toxicity of any substance is measured by its LD50 value, which is equivalent to the amount of the substance needed to kill _____% of a population of animals.

8. GC-MS, HPLC, and TLC are laboratory methods used to identify unknown _____.

9. The type of alcohol that is produced from the fermentation of certain fruits and grains is _____ alcohol.

10. The alcohol referenced in question 9 is sometimes called _____ alcohol.

11. The chemical formula seen here is that for _____ alcohol. **CH_3OH**

12. The minimal blood alcohol level for a DUI in most states is _____ mg %.

13. Opium and morphine fall under the heading in drug classification of being _____.

14. Heroin is prepared from morphine by a synthesis reaction with _____.

15. The lethal limit on the BAC scale for ethyl alcohol is above _____ mg %.

16. LSD, PCP, and mescaline are all classified as _____ agents that significantly alter normal thought processes, perceptions, and moods.

FORENSIC ENTOMOLOGY

LEARNING OBJECTIVES

After completion of this chapter, students will be able to:

- » Explain the significance of the arthropods in the living world
- » Differentiate various life cycles of insect orders
- » Identify various insects in the orders and families important to the forensic entomologist
- » Utilize the insect life cycle as a biological clock to determine PMI
- » Explain the various ways in which forensic entomology may be useful in solving a criminal event
- » Compare the various stages that occur in human decomposition

Entomology is the science that deals with the study of insects. Insects are animals that are included in the larger group of invertebrates called arthropods. The arthropods make up the largest of all the phyla in the animal kingdom and include the classes of insects, crustaceans, arachnids, millipedes and centipedes. More than 75% of all the animals on earth are found in this group called the arthropods. Within the five classes mentioned above the largest is the class of insects which accounts for approximately 70%-80% of all of the arthropods.

Insects are found in just about every niche on the globe with the exceptions of the most distant Polar Regions. There are several million species of insects in the world and more are being discovered every year. Their habitats are varied and their life cycles are well defined. The application of the study of insects as they are related to medical or legal issues provides the basis for the field of forensic entomology. An understanding of the ubiquitous nature and biology of insects is helpful in developing the concepts that allow the forensic entomologist to apply the study of entomology within the realm of medical and legal issues. Forensic entomology might apply in civil litigation where insects may be associated with food contamination and the installation of contaminated wood products. A major area of the use of forensic entomology is that of death investigations.

Certain carrion insects have a unique ability to locate decomposing flesh or an open wound. The common blowfly may have the ability to locate decomposing animal material up to two miles away. The insects associated with a decomposing body can tell investigators a great deal about the circumstances surrounding the death. The following areas of interest associated with a decomposing body are:

1. determination of time of death or post mortem interval (PMI)
2. time of year
3. location of body - movement of the corpse
4. manner and cause of death
5. association of suspects with death scene
6. detection of toxins or drugs through analysis of insect larva.

Understanding how the factors mentioned above relate to forensic entomology relies strongly on the following ecological concepts.

1. Different species of insects go through developmental stages, life cycles, in an extremely predictable manner.
2. The time period for the life cycle of a specific insect is temperature dependent at the microhabitat level.
3. Ecological succession occurs which involves a corpse being invaded by a series of different species or insect groups over time. Each species or group changes the microenvironment as a result of its activities which makes it attractive to new waves of organisms.

The most important environmental factors associated with corpse decay, based on a study of decay rates of 150 human corpses at the University of Tennessee, are temperature, access by insects and depth of burial.

There are a number of life cycles that can be described among the various groups of insects. These life cycles follow specific patterns such as those seen in figure 9.1.

Gradual development as is common in insects such as the silverfish goes from the egg to the developing young to the mature adult. Incomplete metamorphosis such as seen in the grasshopper, proceeds from the egg to the nymph to the adult. Complete metamorphosis which is common to a majority of the insects involves changes from the egg stage to the larval development followed by the pupa formation and finally a mature adult. The knowledge of these various life cycles provides the forensic entomologist with a very specific biological clock. The time that it takes for an insect egg to develop into an adult is a well described time period which has been

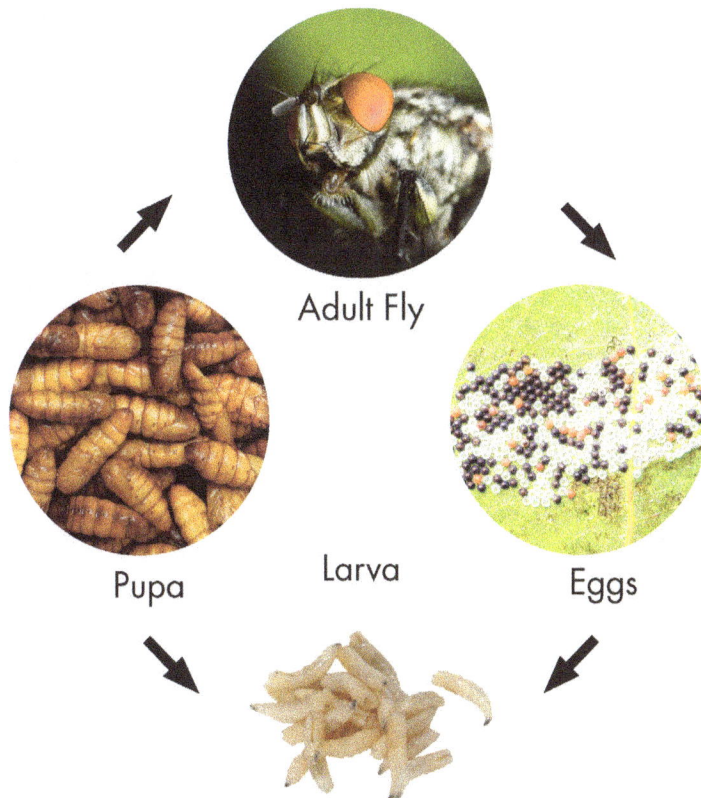

Adult Fly

Pupa   Larva   Eggs

FIGURE 9. 1A-D
Life cycle of a blow fly.

documented for most all of the insect species. Knowing the life cycle and being able to identify the life form changes allows the entomologist to specify the post mortem interval (PMI) or the time since death. The life cycle of the blowfly, seen in figure 9.1, illustrates the changes that take place and establishes a time frame within which these intermediate forms are present. One of the first groups of insects that arrive on a dead vertebrate is usually blowflies (Diptera: Calliphoridae). Usually the female oviposits (lays eggs) within two days after death of the vertebrate. Then the blowfly goes through the following stages: egg, 1st instar larvae, 2nd instar larvae, 3rd instar larvae, prepupae, pupae within puparium, imago. If we know how long it takes to reach the different stages in an insect's life, we can calculate the time since the egg was laid. This calculation of the age of the insects can be considered as an estimate of the time of death.

The two key factors that allow for estimates of the PMI using various insect populations are the time period for a given species to reach a particular stage of development and the comparison of assemblages of insect fauna on the corpse at the time of examination. Insect colonization of a corpse is predictable and flies are the first insects to recruit to a body, often within minutes. Colonization begins in moist body cavities such as the nose or oral cavity and if any types of wounds occur on the body insect colonization will be evident at the site of the wound (figure 9.2).

FIGURE 9. 2
Blowfly infestation.

If insect colonization occurs in the chest region, not a normal open cavity, you might suspect some type of wound in that area such as that associated with stabbings or gunshot wounds. Another important environmental factor which will affect colonization is that of seasonal pattern. In certain locals such as the northern latitudes many insect populations are dormant and colonization may not occur in the same manner and time frame depending on whether it is winter or summer.

We can study the ecological roles of insects in decomposition by describing their nutritional life styles. The necrophages, species feeding on corpse tissue, include mostly the flies and beetles. Their age estimation which is determined by identifying specific larval instar stages and pupa are very important for determining PMI. The omnivores, insects that feed both on the corpse and associated fauna such as ants, wasps and some other beetles may alter the rate of decomposition. Parasites and predators including many of the beetles, true flies and wasps may attack immature flies and alter the time period between egg laying and larval maturation. Finally there are arthropods called incidentals which may use the corpse as a resource extension and this group includes springtails, spiders (arachnids) centipedes and possibly some mites. As you can see the time period for decomposition by insects is an extremely complex and intertwined process.

The stages of decomposition occur with mostly predictable changes over time. We can study the decompositional changes in five stages as listed below and seen in figure 9.3.

Stage 1: Initial decay 0-3 days after death
Stage 2: Putrefaction 4-10 days after death
Stage 3: Black putrefaction 10-20 days after death
Stage 4: Butyric fermentation 20-50 days after death
Stage 5: Dry decay 50-365 days after death.

Stage 1, initial decay, begins at the moment of death as many species of flies are attracted to the corpse. The flesh flies, family Sarcophagidae and the blowflies' family Calliphoridae are the first insects to arrive. The flies in figure 9.4 enter the body orifices (nose, oral cavity, anus, and vagina) or locations of wounds on the body and lay eggs. After 24-48 hours these eggs hatch into first instar larva and feed on the flesh. During this time the body, lacking any of its internal defenses, begins to decompose internally as a result of bacterial decomposition which usually begins in the intestine and parts of the GI tract. The larval stages of development may take anywhere from 2-4 weeks depending on the insect species and other environmental factors such as temperature, wind and rain.

Stage 2, putrefaction, begins with production of gases by the action of anaerobic bacteria in the GI tract which eventually results in the seepage of fluids from internal body cavities. At this time larval decomposition speeds up and the odors and body fluids that begin to be released from the body attract more blowflies, flesh flies, beetles and mites. The later arriving flies and beetles are predators, feeding on the larva (maggots) present as well as decaying flesh.

Additionally parasitoid wasps lay their eggs inside the fly larva and later inside the pupa; which have formed from the developing larva.

Stage 3, black putrefaction, occurs when the abdominal wall is broken and the body deflates as a result of the escaping gases from the internal decomposition. By this time several generations of larva are present on the body and some have become fully grown. They migrate to

FIGURE 9.3
Mice carcass in the different stages of decomposition.

FIGURE 9.4
Insects involved in the decompostion stages.

surrounding areas of the body often in the soil or grass and become pupae. During this time predatory maggots are abundant and retard the arrival of many of the pioneer flies. Predatory beetles lay their eggs on the corpse and upon hatching out their larva begin eating the decomposing flesh. Parasitoid wasps are much more prevalent and continue to lay their eggs inside the larva and pupae on the corpse.

Stage 4, butyric fermentation is the stage when all the remaining flesh on the body is removed and the body begins to dry out. The body has a cheesy smell, caused by the production of butyric acid, and this causes a new influx of corpse organisms. The reduction of soft flesh makes the body less appealing to the hooked mouths of maggots and more suitable to the chewing mouthparts of beetles. Beetles feed on the skin and cartilaginous materials left on the corpse and many of them are larva that hatched out from earlier beetle populations.

Stage 5, dry decay stage finds the body void of moisture and the rate of decay is extremely slow. Eventually the hair disappears and only the skeletal remains are present.

Animals which can feed on hair include certain moths and microorganisms such as bacteria and fungi. Mites in turn feed on these microorganisms. These organisms will remain on the corpse as long as traces of hair are present so that depends on the amount of hair on the animal that is decomposing. Humans and pigs have relatively little hair so this stage is considerably shorter for these particular species.

The following groups of insect orders and families represent many of the insects that are involved in the stages of decomposition:

Order Diptera
FAMILY

- Sarcophagidae   Flesh flies
- Calliphoridae   Blowflies,bottle flies & screwsorm flies
- Muscidae   houseflies
- Phoridae   phorid flies
- Stratiomyidae   soldier flies
- Syrphidae   drone flies
- Piophiladae   cheese skippers

Order Coleoptera – Beetles
FAMILY

- Silphidae   Burrowing beetles, carrion beetles
- Staphylinidae   Rove beetles
- Histeridae   Hister beetles & clown beetles
- Dermestidae   Dermestid or skin beetles.
- Scarabaeidae   Hide beetles

Order Hymenoptera- Bees, ants & wasps

- A variety of different ant species, red ants, fire ants, big headed ants may slow down colonization by removing fly eggs and larvae. Some ant feeding may resemble chemical burns.
- Several wasp species will feed on flies and fly larvae.

Estimating time since death with Forensic Entomology

After the initial decay and the body begins to smell, different types of insects are attracted to the body. The insects that usually arrive first are the Diptera, in particular the blow flies (Calliphoridae) and the flesh flies (sarcophagidae).

The females will lay their eggs on the body, especially around the natural orifices such as the nose, mouth, ears, anus and vagina. If the body has wounds the eggs are also laid around or within the site of the wound.

After some short time, depending on the species, the eggs hatch into small 1st instar larvae. These larvae feed on the dead tissue and grow quite rapidly into 2nd and 3rd stage instar larvae (figure 9.5).

When the larvae are full grown they begin to wander and go into the prepupal stage, followed shortly by the puparium. It typically takes between two to three weeks for most insects to complete this cycle which is largely dependent on the surrounding environmental conditions. The theory behind estimating the PMI is based on the examination of the various stages in the

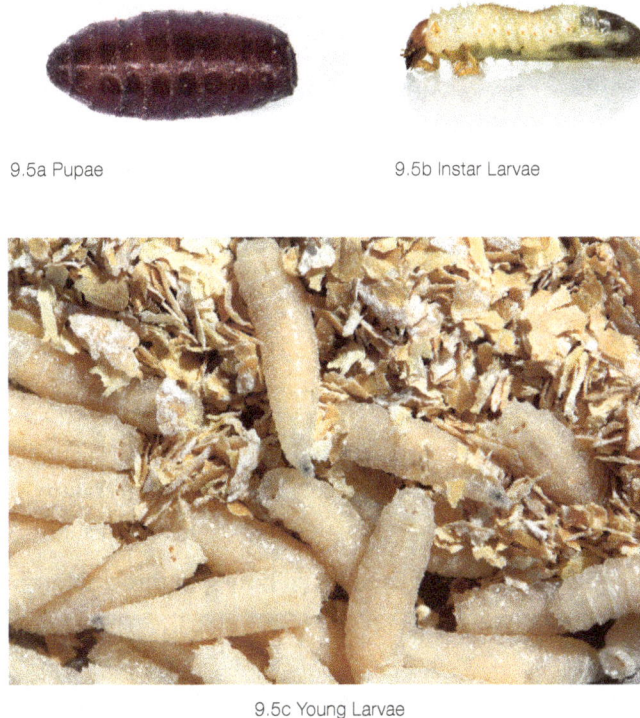

9.5a Pupae

9.5b Instar Larvae

9.5c Young Larvae

FIGURE 9.5
Larvae and pupal stages.

FIGURE 9.6
Lucilia sericata.

metamorphic development and knowing the time periods associated with each of the stages for the various insect populations.

Case study:

ESTABLISH THE PMI:

Body found and all larvae collected are *Lucilia sericata*
The largest and oldest are mid 3rd instar larvae
Weather reports from the crime scene indicate the average temperature is 21^0 C.

Table 9.1 **Development rate of Lucilia sericata (in hours) at three different temperatures.**

Temp °C	Egg	1st instar	2nd instar	3rd instar	Pre-pupa	Pupa	Total time days
18	41	33	42	58	148	383	29
21	21	31	26	50	118	240	24
27	13	20	12	40	80	158	14

SOLUTION

1. Use 21^0 C as the average temperature
2. It takes 21 hours to hatch from the egg, plus 31 hours for 1st instar, plus 26 hours
3. for the 2nd instar, plus 25 hours roughly for the mid-3rd instar larva.
4. Add 48 hours for colonization time.

Total: 21 + 31 + 26 + 25 + 48 = 151 hours

151 hrs/24 hrs = 6.3 roughly 6 days PMI

FOR FURTHER READING

Van Horn, C. (June 14, 2011). Bug expert says coffin flies indicate human decomposition in Casey Anthony's car. *Examiner.com*. Retrieved from http://www.examiner.com/article/bug-expert-says-coffin-flies-indicate-human-decomposition-casey-anthony-s-car

1. _____ is the science that deals with the study of insects.

2. The animal phylum which contains over 60% of all animals in the world is named _____.

3. The forensic entomologist uses the life cycle of various insects to create a biological _____.

4. In the life cycle of a blowfly the stage that follows the egg is the _____ stage.

5. Those flies in the family sarcophagidae are commonly known as the _____ flies.

6. After death, the common blowfly may have the ability to detect a decomposing body from as far as _____ miles away.

7. Insect colonization on a corpse begins in the body's _____ or at the sight of a wound.

8. Because after death our body's defense mechanisms shut down, we begin to decompose as a result of _____ decomposition which usually begins in the intestine.

9. The decompositional stage of _____ occurs when the abdominal wall is broken and the body deflates as a result of escaping gases.

10. The PMI may be determined by measuring the length of the larval stages and equating this with the time of the _____.

11. The insects in the family Piophilidae are commonly known as the _____.

12. The insects commonly known as beetles are classified in the order _____.

FORENSIC BOTANY

LEARNING OBJECTIVES

After completion of this chapter, students will be able to:

» Identify the major components of a typical plant cell
» Explain the major tissues found in a majority of plants
» Discuss the significance of biological succession at a crime scene
» Interpret the findings of various plant parts on a suspect at a crime scene
» Differentiate various plant materials such as wood chips, flower parts, and pollen grains
» Explain how different varieties of plant parts may be significant in a criminal investigation

As Forensic Science is an application of a scientific discipline to the law, Forensic Botany is the application of the plant sciences to legal matters. In many cases this utilizes the knowledge of botanical science, that is, the understanding of plant anatomy and physiology to assist in the solution of crimes such as homicide, kidnapping and to help determine the time of death or PMI (Graham, 1997). Many areas of botany are employed such as plant anatomy, the study of plant cells, plant taxonomy (plant identification), plant systematics (plant relationships to other plants), plant ecology (relationship between plants and their environments) and palynology (the study of plant pollen and spores).

The identification of plant parts such as seeds, flowers, pollen grains, leaves whether occurring on land plants, fresh water or marine plants may enable an investigator to identify the location where a crime was committed or possibly provide the identification of an assailant involved in a criminal offense. The widespread distribution of the many plant species in all environments allows a trained forensic botanist to reference specific plant parts to a geographical location. Very often individuals who visit these outside locations come in contact with these plants and materials are transferred from the plant onto the clothing, footwear or skin of an individual. Careful observation of the clothing or any material that a suspect may have contacted has the possibility of containing these types of plant materials. A thorough knowledge of diversity of plant species and their life cycles enables an examination by a forensic botanist to look for specific plant parts which might give evidence of location, time of year and type of environment.

Most of the time when we think of a plant we think of one of the vascular plants usually one of the flowering plants. The flowering plants belong to a taxonomic division called the Anthophyta, plants whose seeds are encased in an ovary. The angiosperms, gymnosperms, cycads, and ginkgo belong to a larger group of plants, the vascular plants along with the ferns, horsetails and club mosses. Let's consider the flowering plants which are classified based on the number of seed leaves (cotyledons) they have. The monocotyledons are vascular plants which have one cotyledon and are known as monocots. Corn is an excellent example of a vascular monocot. Plants that are dicotyledons, or dicots have two seed leaves such as the bean plant. There a many additional characteristics which help differentiate the monocots and dicots such as leaf venation, specialized tissues in the stems of the plants, as well as characteristics of the floral arrangements. Flowering plants can also be classified as annuals, those that live for only a single growing season or perennials, those that may live for multiple growing seasons. Flowering plants may be considered to be herbaceous, having soft stems, or woody, having harder stems.

In order to develop a better understanding of the gross structure of many plants it is useful to become familiar with the typical cellular structure of plants. A representative plant cell is pictured in figure 10.1.

Plant cells contain cell walls which are largely composed of cellulose, a complex carbohydrate that is very resistant to chemical breakdown. Plant cell walls may remain intact for thousands of years whereas the contents of the plant cell cytoplasm will have broken down at the early

PLANT CELL

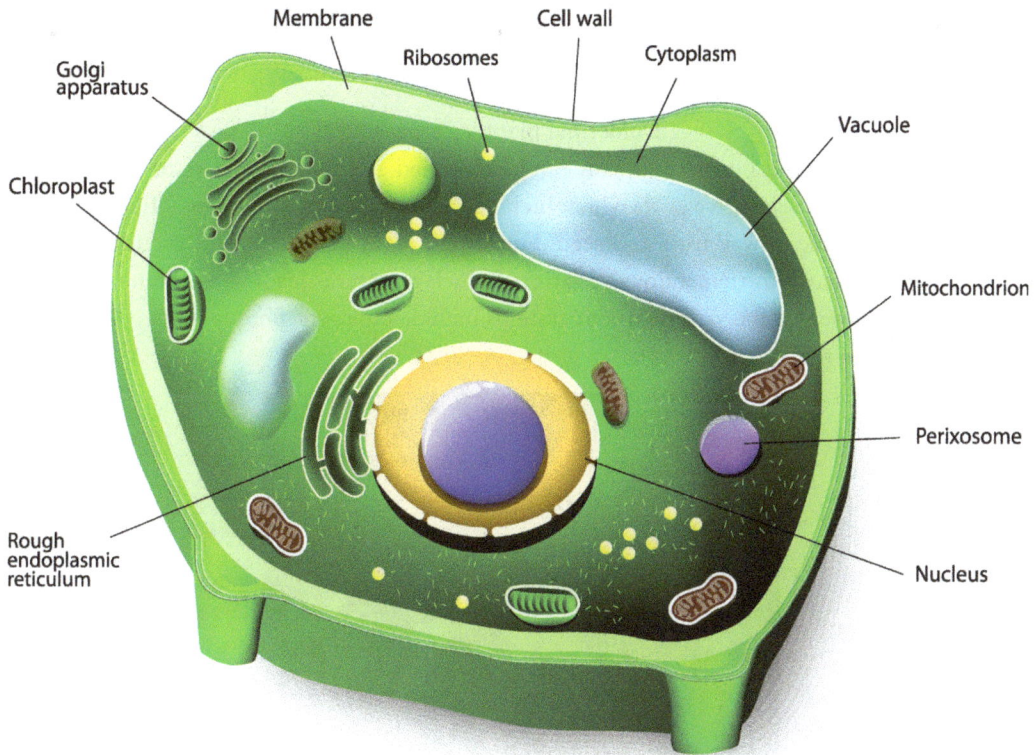

FIGURE 10.1
Plant cell.

stages of decomposition (Bock, 1997). Likewise many forms of pollen grains and spores are composed of materials which are resistant to decay. The identification of pollen grains in the intestine of the "Iceman" a 5000 year old human, frozen in an ice block in the Alps, enabled forensic investigators to track the location at different altitudes where he had traversed the mountainside from lower to higher elevations coming in contact with the pollen of the various species of plants growing at these altitudes.

Not all plant cells contain all of the organelles seen in the diagram above. The differences in size, shape and specific features allow plant cells to differentiate to perform specific functions, such as photosynthesis, absorption and secretion. In addition these differences in cell morphology enable a botanist to identify the various types of plant tissues which are composed of these specific cell types.

The three tissue systems of most plants

Ground tissue system   Metabolic activity, photosynthesis and respiration

                              Parenchyma   Sclerenchyma   Collenchyma

Vascular tissue system   Transportation of materials throughout the plant

                              Xylem and phloem

Dermal tissue system   Protection

                              Epidermal plant surfaces

Plant parts of a typical flowering plant can be classified as either vegetative or reproductive parts. There are three vegetative parts which include the roots, stems and leaves. The reproductive parts include the flower, fruit and seed. Within all plant groups these structures may be modified which results in tremendous variation of such structures leading to the multitude of different plant varieties. The variety of root systems include tap roots, fibrous roots, adventitious roots, and storage roots (figure 10.2).

FIGURE 10.2A-C

fibrous roots, adventitious roots, and storage roots.

The variety of stem types include herbaceous and woody stems (figure 10.3). Herbaceous stems are those that lack secondary growth and are somewhat flexible. In temperate climates these stems usually only last for one growing season. If the plant does overwinter new stems will grow during the next growing season which is usually in the spring. Woody stems exhibit secondary growth and become extremely rigid due to the accumulation of xylem. A woody stem will remain for many years and most have structures which are useful in plant identification such as: terminal buds, bud scales, leaf scars and lenticels. In cross section woody stems display distinct regions and layers which allow the botanist to detect seasonal growth, age, and plant species (Townsley, 2001).

Leaves are the primary sites of photosynthesis and typical leaf structure includes a broad flattened blade with a pattern of venation that allows for the conduction of water and nutrients and a petiole which attaches the leaf to the plant (figure 10.4).

The variation in leaf design, size and shape is related to the tips of the leaves as well as the margin and the manner in which the leaf is attached to the petiole and a variety of other characteristics which become useful in plant identification. Figure 10.6 is a reference to a large variety of leaf characteristics.

The reproductive structures of a plant include the flower, fruit and the seeds. All of these structures become useful in the identification of a particular plant species. The typical flower consists of the sepals, collectively called the calyx, the petals collectively called the corolla, the stamen which consists of an anther and a filament and the pistil which consists of the stigma, style and the ovary. These parts of the flower are identified in figure 10.5.

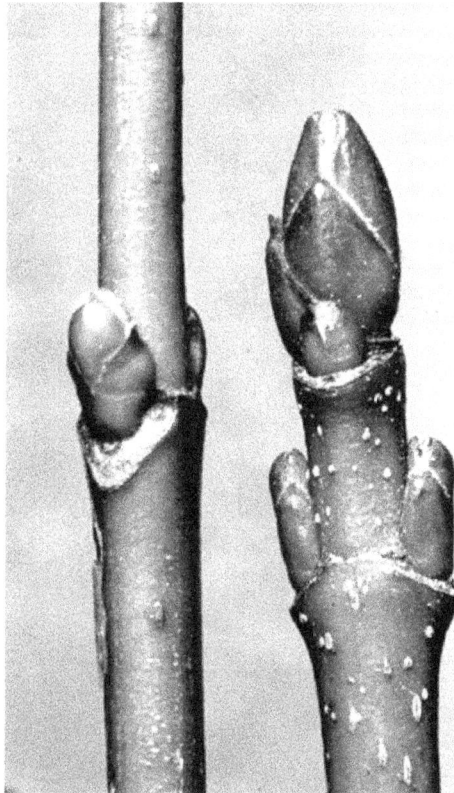

FIGURE 10.3
Woody stem.

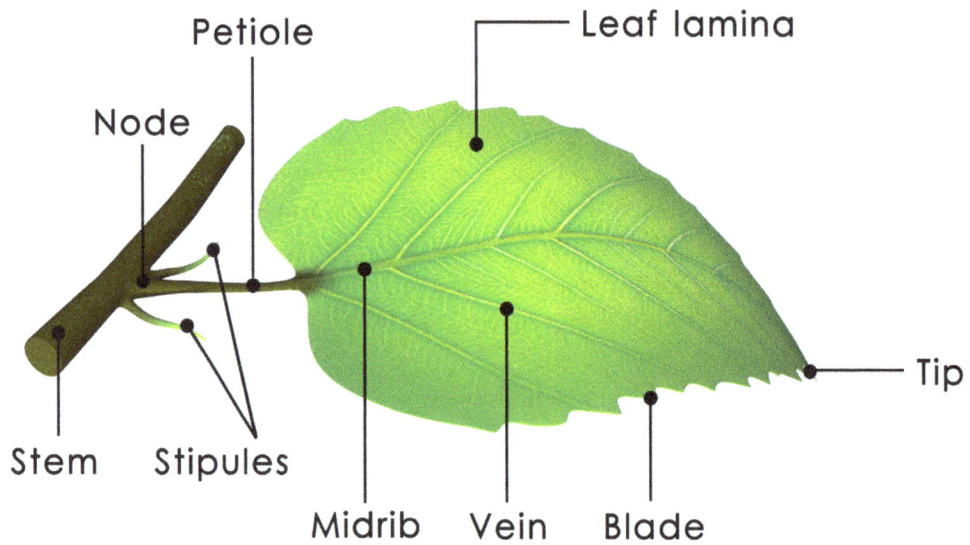

FIGURE 10. 4
Parts of a leaf.

Parts of a Flower

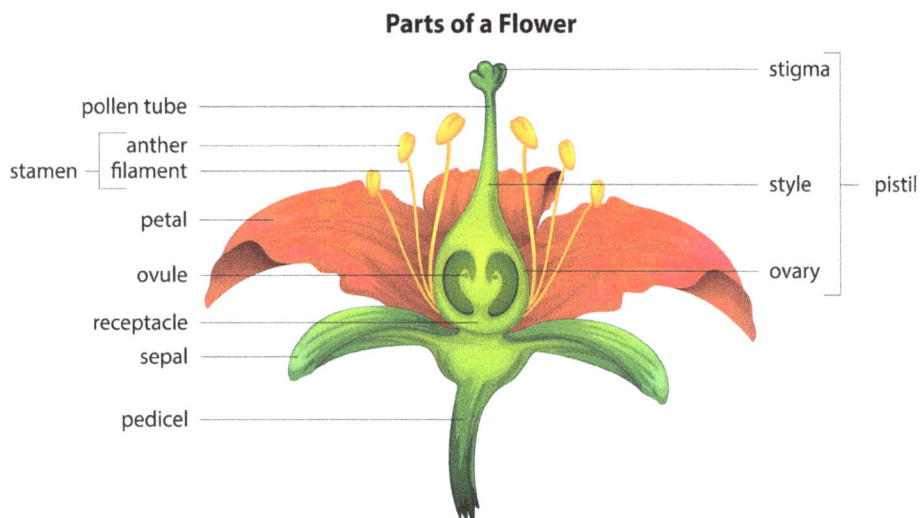

FIGURE 10. 5
Parts of a flower.

LEAF TYPES

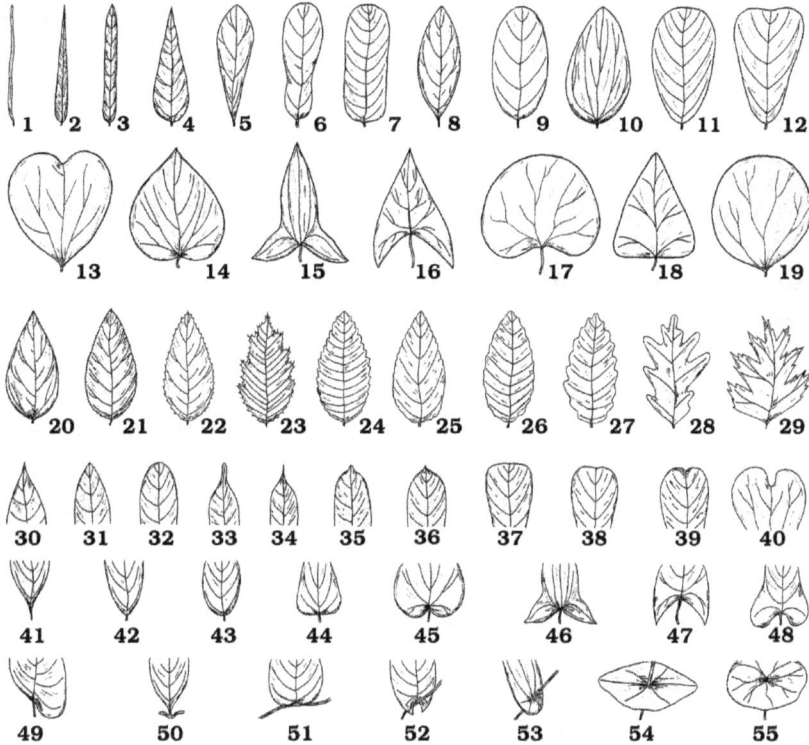

SHAPES
1. Acicular
2. Subulate
3. Linear
4. Lanceolate
5. Oblanceolate
6. Spatulate
7. Oblong
8. Elliptical
9. Oval
10. Ovate
11. Obovate
12. Cuneate
13. Obcordate
14. Cordate
15. Hastate
16. Sagittate
17. Reniform
18. Deltoid
19. Orbiculate

MARGINS
20. Entire
21. Serrulate
22. Serrate
23. Doubly serrate
24. Dentrate
25. Crenate
26. Undulate
27. Sinuate
28. Lobed
29. Incised

TIPS
30. Acuminate
31. Acute
32. Obtuse
33. Caudate
34. Aristate
35. Cuspidate
36. Mucronate
37. Truncate
38. Retuse
39. Emarginate
40. Obcordate

BASES
41. Acuminate
42. Acute
43. Rounded
44. Truncate
45. Cordate
46. Hastate
47. Sagittate
48. Auriculate
49. Oblique
50. Stipulate
51. Sessile
52. Clasping
53. Perfoliate
54. Connate
55. Peltate

FIGURE 10.6

Leaf types.

A fruit is a matured ripened ovary and contains the fertilized ovules that develop into seeds. There is a large variety of fruit types each having unique characteristics which enable the botanist to identify the fruit as well as the plant from which it came. The common names given to fruits are very much different than the botanical names. The figure 10.7 shows many of the fruit types and includes the common and botanical names.

A Seed is a matured ripened ovule. The seed contains the embryonic structures which enable it to develop into a new plant. These structures include the embryonic root, shoot and leaves. The embryo is surrounded by the endosperm and the cotyledons. The endosperm protects the developing embryo and the cotyledons provide nutritional materials to the developing plant. This entire structure is surrounded by a seed coat which offers protection and prevents drying out. Many seeds may lie dormant for years until the favorable environmental conditions of temperature, moisture and light are adequate for the growth and development of the seed. Figure 10.8 shows the seed and structures of a typical monocot and dicot plant.

The understanding of the science of botany and the knowledge of the structure and function of plants may help in a forensic investigation of a homicide or in the reconstruction of the events that led to a crime. Perhaps the most famous use of forensic botany was the botanical evidence that led to a conviction in the Lindberg kidnapping. In this case Dr. Koehler, an expert in plant anatomy, was able to match the wood in the ladder that was used to climb to the second floor of the Lindberg home with the wood in the attic floor of the home of Bruno Hauptman the convicted kidnapper. Since that trial in 1935, what was termed forensic botany, the use of plant remains to help solve crimes or aid in other legal matters has been widely accepted as valid scientific evidence in a court of law (Graham, 1997).

Forensic botany is a new and expanding field. Many criminal investigators, medical examiners and attorneys are not aware of its forensic value mainly due to the lack of exposure to the science of botany in their educational experiences and their career fields. Most forensic botanists perform their work as private consultants in matters dealing with criminal investigations. To be accepted to testify in a court case a forensic botanist must demonstrate that they are qualified to present expert testimony. Their suitability for such testimony is judged by their experience and educational credentials (Lane, 1990).

FOR FURTHER READING

Burstein, J. (2011, January 31). CSI with a botany degree: Plants can help solve crimes. *Sun Sentinel*. Retrieved from http://articles.sun-sentinel.com/2011-01-31/news/ fl-forensic-botany-20110126_1_forensic-botany-indoor-plants-botanist

FRUIT TYPES

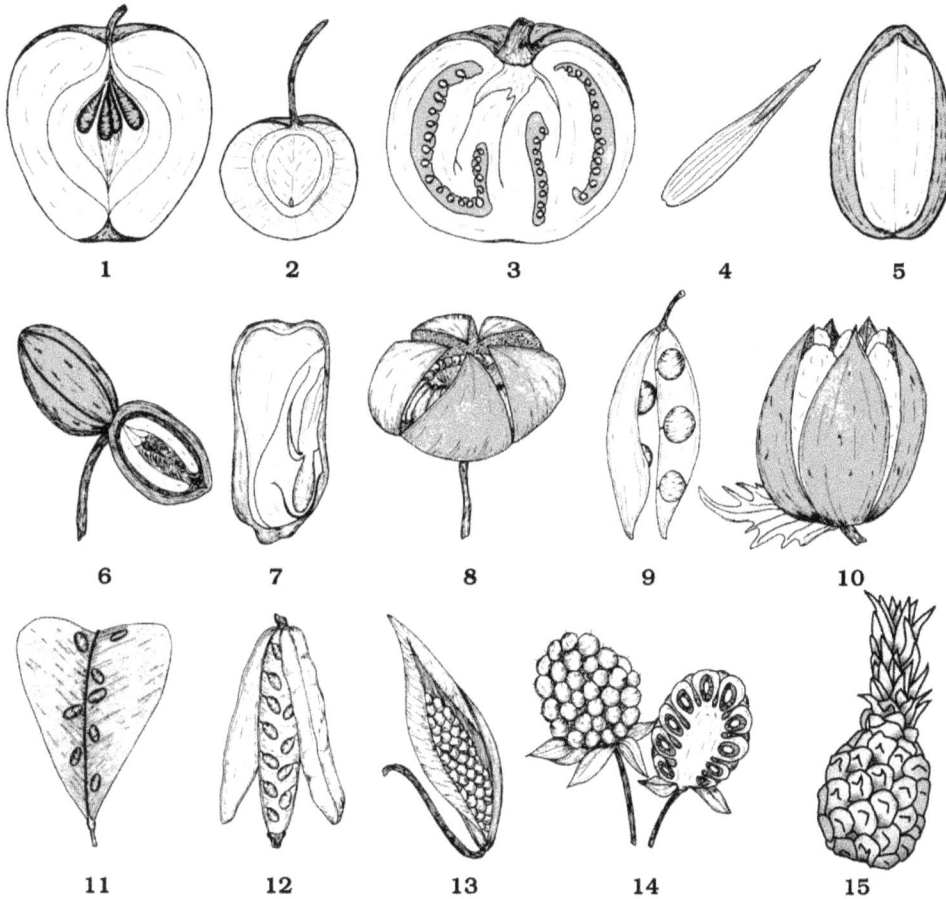

SIMPLE FRUIT
Fleshy
1. Pome (apple)
2. Drupe (cherry)
3. Berry (tomato)
Dry, indehiscent
4. Samara (ash)
5. Achene (sunflower)
6. Nut (pecan)
7. Grain (corn)
8. Schizocarp (geranium)

SIMPLE FRUIT
Dry, dehiscent
9. Legume (pea)
10. Capsule (cotton)
11. Silicle (shepherd's purse)
12. Silique (mustard)
13. Follicle (milkweed)
COMPOUND FRUIT
14. Aggregate fruit (blackberry)
15. Multiple fruit (Pineapple)

FIGURE 10.7

Fruit types.

1. The three vegetative parts of most plants are the _____, _____, and _____.
2. The flower, fruit, and seed make up the _____ parts of a plant.
3. Plants that have soft stems that lack secondary growth are said to be _____.
4. In land plants, the _____ system helps anchor the plant in the soil.
5. The primary sites of _____ are found in the leaves of the plant.
6. In flowering plants, the pollen grains are disseminated from the _____, which is the male reproductive organ of the plant.
7. The fruit that we eat is actually a matured ripened ovary, which contains the _____ that may develop into a new plant.
8. The tremendous diversity in the plant population may allow a forensic botanist to determine the _____ of a crime by analyzing things such as pollen, leaf parts, and seeds.
9. Plant cells contain _____ in their cell walls, which is extremely resistant to decomposition.
10. Pollen grains in the intestine of the "Iceman" enabled forensic scientists to track the various _____ that he had been traversing.

REFERENCES

Adkins Bee Removal—Yellow Jackets Picture 1. (2014). Retrieved from http://www.adkinsbeeremoval.com/bee-id-chart/yellowjacket1.html

Alberts, B., Johnson, A., Lewis, J., Raff, M., Roberts, K., & Walter, P. (2002). Molecular biology of the cell (4th ed.). New York: Garland Science. Figure 1-2, DNA and its building blocks. Retrieved from http://www.ncbi.nlm.nih.gov/books/NBK26864/figure/A8

Anderson, D. (n.d.). Electrophoresis. Retrieved from http://www.occc.edu/biologylabs/Documents/Electrophoresis/Question_3.htm

Arizona master gardener manual. (1998). Retrieved from http://ag.arizona.edu/pubs/garden/mg/index.html

Baumgarten, A. (2014). Antigen-antibody reaction. In *AccessScience*. McGraw-Hill Education. Retrieved from http://accessscience.com/content/antigen-antibody-reaction/040700

Bevel, T., & Gardner, R. (2002). *Bloodstain pattern analysis with an introduction to crime scene reconstruction* (2nd ed.). Boca Raton, FL: CRC Press.

Bloodstain pattern analysis. (2013). Retrieved from http://www.neiai.org/neiai/uploads/Kent Smotherman/files/Bloodstain Pattern Analysis.pdf

Bloodstain pattern analysis terminology: SWGSTAIN. (2008). Retrieved from http://hemospat.com/terminology/index.php?cat=projected&sub=MVIS

Blowfly image. (n.d.). Retrieved from http://wikis.lib.ncsu.edu/images/f/f9/Blowflies_3.jpg

Bock, J. H. and D. O. Norris, 1997, *Forensic botany: An under-utilized resource, Journal of Forensic Sciences,* 42, 364-7.

Bowers, M. (2011, October 23). Ted Bundy Bitemarks and Richard Milone: How DNA, bitemark research and failed cases have changed bitemark analysis. *Forensic Odontology—Bitemark Evidence*. Retrieved from http://bitemarks.org/2011/10/23/ted-bundy-bitemarks-and-richard-milone-how-dna-bitemark-research-and-failed-cases-have-changed-bitemark-analysis

Buck, U., Kneubuehl, B., Näther, S., Albertini, N., Schmidt, L., & Thali, M. (2011). 3D bloodstain pattern analysis: Ballistic reconstruction of the trajectories of blood drops and determination of the centres of origin of the bloodstains. *Forensic Science International*, 206(1–3), 22–28.

Burn classification. (n.d.). Retrieved from http://hospitals.unm.edu/burn/classification.shtml

Burn recovery. (n.d.). Retrieved from http://www.burn-recovery.org/injuries.htm

Campbell, N. (2004). Biology: Exploring life, online textbook. Retrieved from http://mtchs.org/BIO/text/chapter27/concept27.1.html

Carolina LabSheets. (2014). *Carolina LabSheets*. Carolina Biological Supply Company. Retrieved from http://www.carolina.com/teacher-resources/carolina-labsheets/21801.co.

Carter, P. (n.d.). Chapter 6: Bones and skeletal tissues. Retrieved from http://classes.midlandstech.edu/carterp/Courses/bio210/chap06/lecture1.html

Clark, K., Evans, L., & Wall, R. (2006). Growth rates of the blowfly, *Lucilia sericata*, on different body tissues. *Forensic Science International*, 156(2–3), 145–149.

Common green bottle fly—*Lucilia sericata* (Meigen). (2011). Retrieved from http://entnemdept.ufl.edu/creatures/livestock/flies/lucilia_sericata.htm

Conjunctival Petechiae. (n.d.). Retrieved from http://medicalpicturesinfo.com/conjunctival-petechiae

Darwin's Dachshund. (2011, March 25). Why are calico cats always female? *Unzip Your Genes* [blog]. Retrieved from https://unzipyourgenes.wordpress.com/2011/03/25/why-are-calico-cats-always-female

Dean, L. (2005). Blood groups and red cell antigens [Internet]. Bethesda, MD: National Center for Biotechnology Information. Chapter 2: Blood group antigens are surface markers on the red blood cell membrane. Retrieved from http://www.ncbi.nlm.nih.gov/books/NBK2264

Dean, L. (2005). Blood groups and red cell antigens [Internet]. Bethesda, MD: National Center for Biotechnology Information. Chapter 5: The ABO blood group. Retrieved from http://www.ncbi.nlm.nih.gov/books/NBK2267

Essig, F. (2012, March 28). The "root" of the root problem. *Botany Professor* [blog]. Retrieved from http://botany-professor.blogspot.com/2012_03_01_archive.html

Fingerprint line types. (2009, May 28). *Behind the Crime: Investigative Tools of Crime to Conviction* [blog]. Retrieved from https://behindthecrime.wordpress.com/about/fingerprints/fingerprint-line-types

Fischer, H., Hammel, P., & Dragovic, L. (2003). Human bites versus dog bites. *New England Journal of Medicine*, 349(11), e11.

Freeman, S. (2008, April 24). How bloodstain pattern analysis works. *HowStuffWorks Science*. Retrieved from http://science.howstuffworks.com/bloodstain-pattern-analysis.htm

Garcia, I. (2010). Biology final exam, flashcards. Retrieved from https://quizlet.com/2428335/biology-final-exam-flash-cards

Gardner, R. M. (2004). *Practical crime scene processing and investigation*. Boca Raton, FL: CRC Press.

Garg, V., Oberoi, S., Gorea, R., & Kaur, K. (2004). Changes in the levels of vitrous potassium with increasing time since death. *JIAFM*, 26(4), 136–139. Retrieved from http://forensicwayout.com/Portals/0/vitreous K.pdf

Geberth, V. (2007). Bloodstain pattern analysis. Retrieved from http://www.practicalhomicide.com/Research/LOmar2007-2.htm

Girard, J. (2008). *Criminalistics: Forensic science and crime*. Sudbury, MA: Jones and Bartlett.

Gleeson, M. (n.d.). Hair growth—Hair removal. Retrieved from http://www.thehazelsbeauty.net/advancedelectrolysis.htm

Gun shot wound to hand. (n.d.). Retrieved from http://s46.photobucket.com/user/MedFX_photos/media/GSW-hand-entrance-and-exit.jpg.html

Hoovler, E. (2012, May 25). 25 pictures of the human body under an electron microscope. *Alizul: A Blog of Informative Contents* [blog]. Retrieved from http://alizul2.blogspot.com/2012/06/25-extraordinary-images-of-human-body.html

Human fetal skulls set of 12. (n.d.). Bone Clones, Inc.—Osteological Reproductions [website]. Retrieved from https://boneclones.com/product/human-fetal-skulls-set-of-12

Jackson, A., & Jackson, J. (2004). *Forensic science*. Harlow, UK: Pearson Prentice Hall.

Lamb, R. (2008, April 29). Housefly life cycle. *HowStuffWorks Animals*. Retrieved from http://animals.howstuffworks.com/insects/housefly4.htm

Lane, M.A., L.C. Anderson, T.M. Barkley, J.H. Bock, E.M. Gifford, D.W. Hall, D.O. Norris, T.L. Rost, and W.L. Stern. 1990. Forensic botany: plants, perpetrators, pests, poisons, and pot. Bioscience 40(1):34-39.

Larvae and pupae stages. (n.d.). Retrieved from http://ars.els-cdn.com/content/image/1-s2.0-S0965174809000356-gr7.jpg

Lazaroff, M., & Rollison, D. (n.d.). Classification of fingerprints. Retrieved from http://shs2.westport.k12.ct.us/forensics/04-fingerprints/classification.htm

Lyle, D. (2013, April 4). Carbon monoxide: A deadly gas. *Writer's Forensics* [blog]. Retrieved from https://writers-forensicsblog.wordpress.com/2013/04/04/carbon-monoxide-a-deadly-gas

Marras, N. (2008). The beginning of the calculus. Retrieved from http://www.nicolamarras.it/calcolatoria/esordi_calcolo_en.html

Mennear, D. (2011, February 28). Skeletal series A: The biological basis of bone and anatomical directional terms. *These Bones of Mine* [blog]. Retrieved from https://thesebonesofmine.wordpress.com/2011/02/28/the-biological-basis-of-bone-anatomical-directional-terms

Plant cell structure. (n.d.). Retrieved from http://en.wikipedia.org/wiki/File:Plant_cell_structure_svg.svg

Poduval, M. (2013). Skeletal system anatomy in children and toddlers (T. Gest, Ed.). Medscape. Retrieved from http://emedicine.medscape.com/article/1899256-overview

Raven, P., & Johnson, G. (2002). *Biology* (6th ed.). Boston: McGraw-Hill.

Rohini. (2009). Information about various medical facts [blog]. Retrieved from http://doctor4help.blogspot.com/2009_08_01_archive.html

Saferstein, R. (2001). *Criminalistics: An introduction to forensic science* (7th ed.). Upper Saddle River, NJ: Prentice Hall.

Schweich, M., & Knüsel, C. (2003). Bio-cultural effects in medieval populations. *Economics & Human Biology*, 1(3), 367–377.

A Simplified Guide to Bloodstain Pattern Analysis. (n.d.). Retrieved from http://www.crime-scene-investigator.net/SimplifiedGuideBloodstainPatterns.pdf

Srebrenica Massacre. (2009). Srebrenica Genocide [blog]. Retrieved from http://srebrenica-genocide.blogspot.com

Stem. (2000). *HowStuffWorks Science*. Retrieved from http://science.howstuffworks.com/dictionary/plant-terms/stem-info.htm

Structure: The Kidney. (2010). Retrieved from http://www.aviva.co.uk/health-insurance/home-of-health/medical-centre/medical-encyclopedia/entry/structure-the-kidney

Taxonomic classification. (2013). Retrieved from http://www.ext.colostate.edu/mg/gardennotes/122.html

Tortora, G., & Derrickson, B. (2009). *Principles of anatomy and physiology* (12th ed.). Hoboken, NJ: Wiley.

Townsley, William, Forensic Science Lab Manual Carolina Biological Supply Company, 2001, Burlington ,NCTrimpe, T. (2008). Fingerprints. Retrieved from http://mhs.marbleheadschools.org/teachers/song/Forensics Downloads/Fingerprint Packet.pdf

Ubelaker, D. (1978). *Human skeletal remains: Excavation, analysis, interpretation*. Chicago: Aldine.

US National Library of Medicine. (2006). Visible proofs: Forensic views of the body: Galleries: Technologies: Life cycle of the black blow fly. Retrieved from http://www.nlm.nih.gov/visibleproofs/galleries/technologies/blowfly.html

Voss, S., Cook, D., & Dadour, I. (2011). Decomposition and insect succession of clothed and unclothed carcasses in Western Australia. *Forensic Science International*, 211(1–3), 67–75.

White, R. (1970). A pictoral review of the major insect orders. Retrieved from http://www.sfu.ca/~fankbone/biol/insecta.html

Winfield, M. (n.d.). What are SNPs? Retrieved from http://www.cerealsdb.uk.net/cerealgenomics/WheatBP/Documents/DOC_What_are_SNPs.php

GLOSSARY

Agarose – a polysaccharide obtained from agar, a gelatinous material derived from certain marine algae, that is the most widely used medium for gel electrophoresis procedures.

Agglutination – the process by which suspended bacteria, cells, or other particles are caused to clump together, commonly used in blood typing and identifying the strength of immunoglobins.

Agranulocytosis – a serious and sometimes fatal illness characterized by a reduction of leucocytes, leading to fever and ulcerations of the mucous membranes.

Algor Mortis – the reduction in body temperature and accompanying loss of skin elasticity that occur after death.

Alkaloid – organic based substances found in plants, including many pharmacologically active substances, such as atropine, caffeine, cocaine, morphine, nicotine, and quinine.

Amnion – the thin membrane covering the fetal side of the placenta and also forms the outer layer of the umbilical cord; filled with amniotic fluid.

Anabolic – referring to anabolism, the building of tissue, as in anabolic steroids; testosterone is an example of a natural anabolic steroid.

Anatomical Pathologist – an expert at determining the pathological condition of every organ in the human body by making microscopic observations of the various tissues from the organs.

Angiosperms – flowering plants.

Anthophyta – comprising flowering plants that produce seeds enclosed in an ovary.

Anthropology – the scientific study of the origin, the behavior, and the physical, social, and cultural developments of humans.

Antibodies – specialized cells of the immune system, which can recognize organisms that invade the body such as bacteria, viruses, and fungi. The antibodies then set off a complex chain of events to kill these foreign invaders.

Antigens – substances that, upon entering the body, trigger the production of antibodies.

Asphyxiation – a state of asphyxia or inability to breathe. Oxygen starvation of tissues, which may cause loss of consciousness, stoppage of breathing, and death without artificial respiration.

Attenuated – alive but weakened; pertaining to the reduction in virulence or toxicity of a microorganism or a drug by weakening it.

Autolysis – spontaneous disintegration of cells or tissues by the action of substances, such as enzymes. It generally occurs in the body after death.

Autopsy – an autopsy is an examination of a body after death.

Carrion – dead and decaying flesh, feeding on such flesh.

Catabolic – a metabolic process in which energy is released through the conversion of complex molecules into simpler ones.

Caustic – destructive to living tissue; capable of burning, corroding, dissolving by chemical reaction.

Chondroblast – a cell that originates from a stem cell and forms **chondrocytes**, known as cartilage cells.

Chondrocyte – cells that make up cartilage and consists mainly of collagen and proteins.

Chromatography – any one of several processes for separating and analyzing various chemical substances; the term literally means color writing and is a tool used to detect and identify certain sugars and amino acids in body fluids.

Coagulate – grouping together of small particles in a solution into larger particles; i.e. blood clotting.

Codominant – of or relating to two alleles of a gene that are both fully expressed in a heterozygote.

Cohesive – of or pertaining to the molecular force within a body or substance acting to unite its parts.

Convergence – the property or manner of approaching a limit, such as a point, line, function, or value.

Cotyledons – a simple embryonic leaf in seed-bearing plants, which upon germination either remains in the seed or emerges and becomes green.

Cycads – any of a various cone-bearing evergreen plants that live in tropical regions, have large feathery leaves, and resemble palm trees.

Cytology – the microscopic study of cells, their origin, structure, function, multiplication, and pathology.

Cytoplasm – the substance between the cell membrane and the nucleus that contains the organelles of the cell, cytoskeleton, and various particles.

Deoxyribonucleic Acid (DNA) – a nucleic acid that carries the genetic information in cells, consisting of two long chains of nucleotides twisted into a double helix.

Dermal – of or relating to the skin or dermis.

Dermatoglyphics – the study of the patterns of ridges of the skin of the fingers, palms, toes, and sole of the feet; used to establish identity by anthropologists and law enforcement.

Desiccation – an excessive loss of moisture, the process of becoming dry.

Dicotyledons – any flowering plant having two embryonic seed leaves, flower parts in fours or fives, and net-veined leaves; includes many herbaceous plants an most families of trees and shrubs.

Diuresis – excretion of an unusually large quantity of urine.

Ectoderm – the outermost of the three primary germ layers of an embryo, from which the epidermis, nervous tissue, and, in vertebrates, sense organs develop.

Electrophoresis – a method of separating substances, especially proteins, and analyzing molecular structure based on the rate of movement of each component.

Embryonic – in an early stage of development; of or relating to an embryo.

Endoderm – innermost of the three primary germ layers of an animal embryo, developing into the gastrointestinal tract, the lungs, and associated structures.

Endonuclease – a nuclease that cuts nucleic acids at interior bonds producing fragments of various sizes.

Entomology – the branch of biology concerned with the study of insects.

Eosinophils – a type of white blood cell containing cytoplasmic granules that are easily stained by the red dye eosin or other acid dyes.

Epithelial – cells that form the epithelial tissue that lines both internal and external surfaces of the body and serves a protective function.

Extrapolating – to infer an unknown from something that is known; to estimate the value outside the observed range.

Forensic Anthropology – relating to the use of science or technology in the investigation and establishment of facts associated with human remains that might be used as evidence in a court of law.

Forensic Autopsy - used in an attempt to establish the cause of death; the manner of death, which identifies the instrument or illness causing the death; and the mode of death, which is the circumstance of the death.

Forensic Botany – the use of the formal study of plants in legal proceedings.

Forensic Entomologist – the study of life cycles of insects that feed on the flesh of the dead, to establish time of death and occasionally identify chemicals present in a person's body at the time of death.

Forensic Pathologist – a highly skilled medical doctor who has been trained in clinical and anatomical pathology.

Forensic Toxicologist – uses the science of poisons, their detection and their effects as applied in legal proceedings in courts of law.

Gamete – one of two reproductive cells having half the number of chromosomes (haploid), male (spermatozoon) and female (oocyte).

Genotype – the complete genetic constitution of an organism or group, as determined by the specific combination and location of the genes on the chromosome.

Ginkgo – a tree native to China having fan-shaped leaves; the female plants bear fleshy fruitlike structures containing edible seeds.

Granulocytic – when a group of white blood cells have granules in the cytoplasm.

Gymnosperms – plants that produce seeds that are not enclosed in a fruit or ovary; most are cone-bearing trees or shrubs.

Hemoglobin – complex protein-iron compound in the blood that carries oxygen to the cells; oxygenated hemoglobin is a bright red color whereas cells prior to oxygenation is darker in color.

Heterozygous – having a contrasting pair of genes for any hereditary characteristic at corresponding positions on the chromosomes of an organism.

Homeostasis – a state of equilibrium, as in an organism or cell, maintained by self-regulating processes. In the kidneys, homeostasis is maintained by regulating the amount of salt and water excreted.

Homozygous – having two likes genes for a hereditary trait such as tallness at corresponding position on the chromosomes.

Hyaline – a type of colorless, transparent substance that appears glassy and pink after being stained.

Interstitial – relating to or situated in the small, narrow spaces between tissues or parts of an organ.

Livor Mortis – discoloration of the skin due to the settling of blood in dependent body parts following death, also know as postmortem lividity.

Lymphocytes – white blood cells produced in the lymph nodes and produce antibodies.

Mandible – a fusion of two halves at the mandibular symphysis to form the lower jaw.

Matrix – the material or tissue between cells in which more specialized structures are embedded.

Maxillae – pair of maxillary bones that are fused together to form the upper jaw.

Megakaryocytes – a large bone marrow cell, with a lobed nucleus, whose cytoplasm is the source of blood platelets.

Meiosis – the process of cell division in sexually reproducing organisms in which a nucleus divides into four daughter nuclei, each containing half the chromosome number of the parent nucleus.

Mesoderm – the middle embryonic germ layer, lying between the ectoderm and the endoderm, from which connective tissue, muscle, bone, and the urogenital and circulatory systems develop.

Metabolize – to undergo metabolism, the breaking down of carbohydrates, proteins, and fats into smaller units; reorganizing units into tissue building blocks or energy sources; also eliminates waste produced by this process.

Metamorphosis – change in the form and habits of an animal during normal development; in insects, for example, the transformation of a maggot into an adult fly.

Monocotyledons – any flowering plant having a single embryonic seed leaf, leaves with parallel veins, and flowers with parts in threes: includes, grasses, lilies, and orchids; also known as monocot.

Osteoblast – cells with one nucleus that are responsible for bone formation; become **osteocytes** when trapped in the bone matrix.

Osteology – the branch of anatomy that deals with the structure and function of bones.

Palynology – the science that studies live and fossil spores, pollen grains, and other microscopic plant structures.

Pathology - the study and understanding of the diseased (abnormal) states of the various organ systems in the human body.

Petechiae – small red or purplish spots as a result of capillary hemorrhages, occur in the mucous membrane and skin.

Petroglyph – a carving or line drawing on rock creating a desired pattern or shape.

Phenolphthalein – a colorless crystalline compound used as an indicator for acid and basic solutions.

Photosynthesis – the process in which green plants and certain other organisms make carbohydrates from carbon dioxide and water in the presence of chlorophyll, using light as energy; normally releases oxygen as a byproduct.

Physiology – the biological study of the functions of living organisms and their parts, and of the physical and chemical factors and processes involved.

Postmortem – relating to or occurring during the period after death.

Protuberances – something that bulges out or projects from its surroundings.

Pubic Symphysis – the midline cartilage joint uniting the left and right pubic bones.

Putrefaction – decomposition of organic matter, especially of proteins, by the use of enzymes resulting in production of foul-smelling matter such as hydrogen sulfide, ammonia, and mercaptans.

Rigor Mortis – the rigid stiffening of skeletal and cardiac muscle 1-7 hours after death; it disappears after 1-6 days, or when decomposition begins; also called postmortem rigidity.

Sclera – the dense, white, fibrous membrane covering all of the eyeball except the cornea.

Sebaceous – relating to or resembling fat or sebum, secreting fat or sebum.

Serology – the science that deals with the properties and reactions of serums, especially blood serum.

Spectroscopy – practice of using a spectroscope to obtain information about substances; to plot the intensity versus wavelength of light emitted or absorbed by a substance, compared to the usual characteristic of the substance.

Spheroid – a body that is shaped like a sphere but is not perfectly round, generated by rotating an ellipse on or about one of its axes.

Subcutaneous – as in the layer of loose connective tissue directly under the skin.

Sudoriferous – producing or secreting sweat.

Taphonomy – the study of decaying organisms over time and how they become fossilized.

Tetranucleotide – a compound of four nucleotides.

Thermophilic – any organism requiring temperatures between 45 – 80 degrees Celsius to survive. Organisms that grow best in a warm environment, such as many bacteria.

Thermoregulation – maintenance of a constant internal body temperature independent from the environmental temperature.

Toxicology – the branch of science concerned with the effects, antidotes, and detection of poisons.

Viscous – having relatively high resistance to flow. The molecules of a viscous fluid slide past each other, the friction between them causes the fluid to flow very slowly.

Zygote – the cell resulting from the union of a male and female gamete, after the completion of fertilization, diploid cell.

APPENDIX A

MEDICAL EXAMINER REPORT
Autopsy Reports

AUTOPSY REPORT 1

OFFICE OF THE MEDICAL EXAMINER

NAME: _____JOHN DOE___   ME#<u>0001234</u>

MEDICAL EXAMINER REPORT
REPORT OF AUTOPSY

OFFICIALS PRESENT AT EXAMINATION

None.

EXTERNAL EXAMINATION

The body is secured in a blue body bag with Medical Examiner seal #0001234

The body is viewed unclothed. The body is that of a normally developed, white male appearing the stated age of 19 years with a body length of 74 inches and body weight of 170 pounds, The body presents a medium build with average nutrition normal hydration and good preservation. Rigor mortis is complete, and lividity is well developed and fixed on the posterior surfaces of the body. The body is cold to touch post refrigeration. Short black hair covers the scalp. The face is unremarkable. There is average   body hair of adult-male-pattern distribution, The eyes are closed and have clear bulbar and palpebral conjunctivas, The irides are brown with white sclerae, There are no cataracts or arcus present. The pupils are equal at 5 millimeters, The orbits appear normal, The nasal cavities are unremarkable with an intact septum. The oral cavity presents natural teeth with fair oral hygiene. The ears are unremarkable with no hemorrhage in the external auditory canals. The neck is rigid due to postmortem changes, and there are no palpable masses, The chest is symmetrical. The abdomen is scaphoid.

The upper and lower extremities are equal and symmetrical and present cyanotic nail beds without clubbing or edema. There are no fractures, deformities or amputations present, The external genitalia present descended testicles and an unremarkable penis, The back reveals dependent lividity with contact pallor. The buttocks are atraumatic, and the anus is intact. The integument is of normal color.

OTHER IDENTIFYING FEATURES

There are identification bands on the ankles.

SCARS

- 1 × ½ inch scor — left shoulder
- 1 × ½ inch scar — left hand

TATTOOS

- Symbol with letters — righ upper arm
- Letters - left wrist

There are no other significant identifying features.

The subcutaneous fat measures 1.5 centimeters and is normally distributed, moist and bright yellow. The musculature of the chest and abdominal area is of normal color and texture.

NECK AND TONGUE

The neck presents an intact hyoid bone as well as the thyroid and cricoid cartilages. The larynx has unremarkable vocal cords and folds that appear widely patent without foreign material. The epiglottis is a characteristic plate-like structure without edema, trauma or pathological lesions. Both the musculature and the vasculature of the anterior neck are unremarkable. The trachea and spine are in the midline, and present no traumatic injuries or pathological lesions, The tongue is unremarkable.

CARDIOVASCULAR SYSTEM

The heart weighs 200 grams and there is no chamber hypertrophy or dilatation. The left ventricular wall is 1.1 centimeters and the right is 0.2 centimeters. The cardiac valves appear unremarkable, The coronary ostia are in the normal anatomical location leading into wideiy patent coronary arteries. Right dominant circulation is present. The endocardial surface is smooth without thrombi or inflammation. Sectioning of the myocardium presents no gross evidence of ischemic changes either of recent or remote origin. The aortic arch, along with the great vessels, appears grossly unremarkable.

RESPIRATORY SYSTEM

The lungs are collapsed and together weigh 410 grams. There are no gross pneumonic lesions or abnormal masses identified. The tracheobronchial tree and pulmonary arterial system are intact and grossly unremarkable, The pleural surfaces are pink and smooth with focal mild anthracosis.

HEPATOBILLARY SYSTEM

The liver weighs 1110 grams and presents a brown, smooth, glistening surface. Focal patchy yellow discoloration, due to mild fatty metamorphosis, is present. On sectioning, the hepatic parenchyma is yellow-brown, homogeneous and congested, The unremarkable gallbladder contains approximately 8 milliliters of greenish bile. There is no cholecystitis or lithiasis, The biliary tree is patent. The pancreas presents a lobulated yellow cut surface without acute or chronic pancreatitis.

HEMOLYMPHATIC SYSTEM

The spleen weighs 100 grams and presents a gray-pink intact capsule and a dark red parenchyma. There is no lymphadenopathy, The thymus gland is involuted.

EVIDENCE OF INJURY

PENETRATING GUNSHOT WOUND OF THE CHEST:

The entrance wound is located on the left chest, 18½ inches below the top of the head, ½ inch to the left of the anterior midline and ¼ inch below the nipple. It consists of a ³⁄₂ inch diameter round entrance defect with soot, ring abrasion, and a 2 × 2 inch area of stippling. This wound is consistent with a wound of entrance of intermediate range.

Further examination demonstrates that the wound track passes directly from front to back and enters the pleural cavity with perforations of the left anterior fifth intercostal space, pericardial sac, right ventricle of the heart, and the right lower lobe of the lung, There is no wound of exit.

Three fragments of projectile are recovered. The lead core is recovered in the pericardial sac behind the right ventricle. Two fragments of the jacket are recovered in thé right pleural cavity behind the right lower lobe of the lung.

The injuries associated with the wound: The entrance wound; perforations of left anterior fifth intercostal space, pericardial sac, right ventricle of the heart, right lower lobe of the lung with approximately 1300 milliliters of blood in the right pleural cavity and 1000 in the left pleural cavity; the collapse of both lungs.

Other injuries: There is a ¼ × ⅛% inch small abrasion on the left fourth finger.

EVIDENCE OF RECENT MEDICAL TREATMENT

There is a cardiac monitor pad on the left flank.

EVIDENCE OF ORGAN AND/OR TISSUE DONATION

None.

INTERNAL EXAMINATION: The following excludes any previously described injuries.

BODY CAVITIES

The peritoneum is congested, smooth, glistening and essentially dry; devoid of adhesions or effusion. There is no scoliosis kyphosis or lordosis present. The left and right diaphragms are in their normal location and appear grossly unremarkable.

GASTROINTESTINAL SYSTEM

The esophagus is intact with normal gastroesophageal junctions and without erosions or varices. The stomach is also normal without gastritis or ulcers, and contains 400 milliliters of gastric fluid with food particles. Loops of small and large bowel appear grossly unremarkable. The appendix is unremarkable.

UROGENITAL SYSTEM

The kidneys weigh 100 grams, and 110 grams, right and left, respectively. On sectioning, the cortex presents a normal thickness above the medulla. The renal columns of Bertin extend between the well-demarcated pyramids and appear unremarkable. The medalla presents normal renal pyramids with unremarkable papillae. The pelvis is of normal size and is lined 'by gray glistening mucosa. There are no calculi. The renal arteries and veins are normal. The ureters are of normal caliber lying in their course within the retroperitoneum and drain into an unremarkable urinary bladder containing approximately 75 milliliters of urine.

The external genitalia present an unremarkable penis without hypospadia, epispadias or phimosis. There are no infectious lesions or tumors noted. The descended testicles are of normal size encased with in an intact and unremarkable scrotal sac. There are no abnormal masses or hernias on palpation. The prostate is of normal size and shape and sectioning presents two normal lateral lobes with a thin median lobe forming the floor of the unremarkable urethra. There are no gross pathological lesions.

ENDOCRINE SYSTEM

The thyroid gland is of normal size and shape and presents two well-defined lobes with a connecting isthmus and a beefy-brown cut surface. There are no goitrous changes or adenomas present. The adrenal glands are of normal size and shape. Sectioning presents no gross pathological lesions.

MUSCULOSKELETAL SYSTEM

The ribs, sternum, clavicles, pelvis and vertebral column have no recent fractures. The muscles are normally formed.

CENTRAL NERVOUS SYSTEM

The scalp is intact without contusions or lacerations. The calvarium is likewise intact without bony abnormalities or fractures. The brain weighs 1400 grams and presents moderate congestion of the leptomeninges. The overlying dura is intact and unremarkable. The cerebral hemispheres reveal a normal gyral pattern with severe global edema. The brainstem and cerebelli are normal in appearance with no evidence of cerebellar tonsillar notching. The circle of Willis is patent and presents no evidence of thrombosis or berry aneurysm. Upon coronal sectioning of the brain, the ventricular system is symmetrical and contains clear cerebrospinal fluid. There are no spaceoccupying lesions present. The spinal cord is not examined.

MICROSCOPIC EXAMINATION: One slide examined on June 20, 2013.

HEART: No diagnostic abnormality.

LUNGS: The partial collapse of tissues.

LIVER: No diagnostic abnormality.

TOXICOLOGY: See separate report from NMS Laboratories.

SB

Age: 19   Race: White   Sex: Male   Date: 5/15/2013

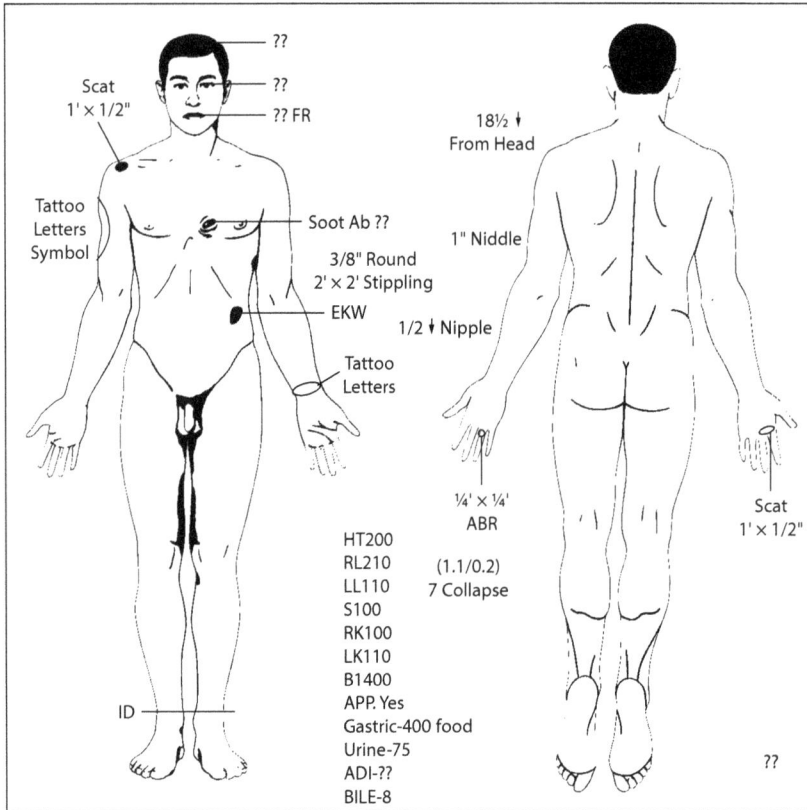

??

??

?? FR

Scat
1' × 1/2"

Tattoo
Letters
Symbol

Soot Ab ??

3/8" Round
2' × 2' Stippling

EKW

Tattoo
Letters

ID

18½ ↓
From Head

1" Niddle

1/2 ↓ Nipple

¼' × ¼'
ABR

(1.1/0.2)
7 Collapse

Scat
1' × 1/2"

??

HT200
RL210
LL110
S100
RK100
LK110
B1400
APP. Yes
Gastric-400 food
Urine-75
ADI-??
BILE-8

Seal # 0001234

Rt Vent
Art Per Card SAC
Proj Post at RT vent, above SAC

| | 1300 | 1000 |

HT $\dfrac{74"}{\text{Fras} \times 2}$ WT $\dfrac{170}{\text{Rt L Lobe ??}}$ BMI $\dfrac{22}{\text{Recovered in cavity}}$ START TIME $\dfrac{11:30}{}$

APPENDIX B

Suggested Laboratories

Chapter 1
Autopsy of the Fetal Pig (Carolina Biological Supply Company)

Chapter 2
ABO Blood Grouping and Paternity Testing (Carolina Biological Supply Company Kit)
Urinalysis and Drug Testing (Carolina Biological Supply Company Kit)

Chapter 3
Blood Spatter Pattern Analysis

Chapter 4
Forensic Anthropology (Study of skeletal remains)

Chapter 5
Forensic odontology, bite mark analysis and dentition

Chapter 6
Fingerprint, hair, fiber analysis (Carolina Biological Supply Company Kits)

Chapter 7

DNA fingerprinting, RLFP, STR analysis, gel electrophoresis (Carolina Biological Supply Company Kits)

Chapter 8

Forensic Toxicology and Drugs (Carolina Biological Supply Company Kits)
Blood Alcohol Testing (Carolina Biological Supply Company Kits)

Chapter 9

Forensic Entomology (The American Biology Teacher volume 63, number 5, May 2003)

Chapter 10.

Forensic Botany (Carolina Biological Supply Company Kit)

APPENDIX C

Answers to Chapter Review Questions

Chapter 1
1. respiratory
2. rigor mortis
3. livor mortis
4. – Y –
5. mid saggital
6. asphyxiation
7. two to four
8. renal or urinary
9. natural, suicide, homicide, accidental
10. petechiae
11. decreased
12. heart
13. contusions
14. puncture

Chapter 2

1. thirty
2. three
3. plasma, red cells and buffy coat
4. antigens
5. B-antigens
6. agglutinate
7. yes
8. female
9. plasma
10. Kastle–Meyer
11. Rh positive
12. 1.5
13. seminal
14. epithelial
15. millimeter

Chapter 3

1. surface
2. fluid
3. five
4. velocity
5. less
6. less
7. direction
8. cast off patterns
9. origin
10. 3-8
11. wiping
12. low
13. skeletonize
14. gushing
15. fibrin clot

Chapter 4

1. identify
2. appendicular
3. osteocyte

4. less than
5. African American or Negroid
6. distal
7. midsaggital
8. female
9. humerus
10. Haversian
11. male
12. 206
13. diaphysis
14. superior
15. decomposition

Chapter 5

1. odontology
2. sixteen
3. deciduous
4. incisor
5. incisor
6. ameloblast
7. identification
8. elastic
9. crown
10. molar

Chapter 6

1. epidermis
2. loops, whorls and arches
3. second
4. cotton, wool, silk
5. rayon, nylon, viscose
6. friction
7. sebaceous
8. cuticle
9. DNA
10. 36
11. fork or bifurcation
12. 4

13. pathognes
14. 1.5 to 2.0

Chapter 7
1. double helix
2. nucleotides
3. thymine
4. PCR
5. amplification
6. deoxyribose
7. enzymes or endonucleases
8. fingerprint
9. hydrogen
10. Watson and Crick
11. chromosomes
12. haploid

Chapter 8
1. toxic
2. intravenous
3. inhalation
4. stimulant
5. hemoglovin
6. hemoglobin
7. 50
8. substances
9. ethyl
10. grain
11. methyl alcohol
12. 0.08
13. narcotics
14. acetic anhydride
15. 0.50
16. hallucinogens

Chapter 9

1. entomology
2. arthropoda
3. clock
4. larvae
5. flesh
6. two
7. orifices
8. bacterial
9. black putrefaction
10. life cycle
11. cheese skipper
12. coleoptera

Chapter 10

1. roots, stems, and leaves
2. reproductive
3. herbaceous
4. root
5. photosynthesis
6. stamen
7. seeds
8. location
9. cellulose
10. altitudes

CREDITS

CHAPTER 3

Fig. 3.1: Copyright © Depositphotos/KrylovVladislav.
Fig. 3.2: Copyright © by Louie Jarvis Photography. Reprinted with permission.
Fig. 3.3: Copyright © by Louie Jarvis Photography. Reprinted with permission.
Fig. 3.4: Copyright © by Louie Jarvis Photography. Reprinted with permission.
Fig. 3.6: Copyright © by Louie Jarvis Photography. Reprinted with permission.
Fig. 3.5: Copyright © by Louie Jarvis Photography. Reprinted with permission.
Fig. 3.8: University of Tennessee, 2003.
Fig. 3.9a: Copyright © by Louie Jarvis Photography. Reprinted with permission.
Fig. 3.9b: Copyright © by Louie Jarvis Photography. Reprinted with permission.
Fig. 3.10: Copyright © by Louie Jarvis Photography. Reprinted with permission.

CHAPTER 4

Fig. 4.2: Copyright © Depositphotos/Lukaves.
Fig. 4.4: Copyright © Depositphotos/stockshoppe.
Fig. 4.5: Copyright © Depositphotos/kniazev.
Fig. 4.6: Copyright © Depositphotos/stihii.
Fig. 4.7: Copyright © Depositphotos/stihii.
Fig. 4.8: Copyright © by Louie Jarvis Photography. Reprinted with permission.
Fig. 4.9: Copyright © by Louie Jarvis Photography. Reprinted with permission.
Fig. 4.10: Copyright © by Louie Jarvis Photography. Reprinted with permission.
Fig. 4.11: Copyright © Depositphotos/boggy22.

CHAPTER 5

Fig. 5.2: Copyright © Depositphotos/eveleen.
Fig. 5.3: Copyright © Depositphotos/roxanabalint.
Fig. 5.4: Copyright © Depositphotos/PicsFive.
Fig. 5.5: Copyright © Depositphotos/simazoran.

CHAPTER 6

Fig. 6.1: Copyright © Depositphotos/stockshoppe.
Fig. 6.2: Copyright © Depositphotos/koratmember.
Fig. 6.3a: Copyright © Depositphotos/jackchen.
Fig. 6.3b: Copyright © Depositphotos/Devon.
Fig. 6.4: Copyright © Depositphotos/PokerMan.

Fig. 6.6: Copyright © Depositphotos/edesignua.
Fig. 6.7: Copyright © Depositphotos/edesignua.
Fig. 6.8a: Copyright © Depositphotos/anele77.
Fig. 6.8b: Copyright © Depositphotos/OlegDoroshenko.
Fig. 6.9: Copyright © Depositphotos/rob3000.

CHAPTER 7

Fig. 7.2: Copyright © Depositphotos/Zerbor.
Fig. 7.1: Copyright © Depositphotos/edesignua.
Fig. 7.3: Copyright © Depositphotos/panuruangjan.
Fig. 7.4: Copyright © Depositphotos/meletver.
Fig. 7.5: Copyright © Depositphotos/edesignua.

CHAPTER 9

Fig. 9.1: Copyright © Depositphotos/Ale-ks.
Fig. 9.1b: Copyright © Depositphotos/kristt.
Fig. 9.1c: Copyright © Depositphotos/zhangyuangeng.
Fig. 9.1d: Copyright © Depositphotos/dusan964.
Fig. 9.1e: Copyright © Depositphotos/jianghongyan.
Fig. 9.2: Copyright © Depositphotos/kolidzeitattoo.
Fig. 9.5a: Copyright © Depositphotos/akova777.
Fig. 9.5b: Copyright © Depositphotos/Kokhanchikov.
Fig. 9.5c: Copyright © Depositphotos/abet.
Fig. 9.6: Copyright © Depositphotos/lifeonwhite.

CHAPTER 10

Fig. 10.1: Copyright © Depositphotos/edesignua.
Fig. 10.2a: Copyright © Depositphotos/madozi.
Fig. 10.2b: Copyright © Depositphotos/Irin717.
Fig. 10.2c: Copyright © Depositphotos/griffin024.
Fig. 10.3: Copyright © by Carolina Biological Supply Company. Reprinted with permission.
Fig. 10.4: Copyright © Depositphotos/sciencepics.
Fig. 10.5: Copyright © Depositphotos/blueringmedia.
Fig. 10.6: Copyright © by Carolina Biological Supply Company. Reprinted with permission.
Fig. 10.7: Copyright © by Carolina Biological Supply Company. Reprinted with permission.

APPENDIX A

Adapted from Medical Examiner Report: Report of Autopsy completed by the Office of the Medical Examiner, State of Florida, Districts 7 & 24, 2012. Names and identifying details have been changed to protect the privacy of individuals.

www.ingramcontent.com/pod-product-compliance
Lightning Source LLC
Chambersburg PA
CBHW081533220326
41598CB00036B/6422